KT-178-353

WITHDRAWN

# The technology of suspended cable net structures

*Frances Chaplin, Geoffrey Calderbank and Jacqueline Howes*

**Construction Press** *London and New York*

THE LIBRARY
GUILDFORD COLLEGE
of Further and Higher Education

102502
721.042 CHA

**General Editor: Colin Bassett, BSc, FCIOB, FFB**

**Construction Press**
*an imprint of:*
Longman Group Limited
Longman House, Burnt Mill, Harlow
Essex CM20 2JE, England
Associated companies throughout the world

*Published in the United States of America*
*by Longman Inc., New York*

© Longman Group Limited 1984

All rights reserved, no part of this publication may be reproduced, stored in a retrieval system, or transmitted in any form or by any means, electronic, mechanical, photocopying, recording, or otherwise, without the prior written permission of the Publisher.

*First published 1984*

**British Library Cataloguing in Publication Data**
Chaplin, Frances
The technology of suspended cable net structures.
1. Cables – Design and construction
2. Bridges, Suspension – Design and construction
I. Title II. Calderbank, Geoffrey
III. Howes, Jacqueline
624.1'774 TA660.C3

ISBN 0-86095-040-9

**Library of Congress Cataloging in Publication Data**
Chaplin, Frances, 1955-
The technology of suspended cable net structures.

Bibliography: p.
1. Cable structures. I. Calderbank, Geoffrey, 1945- II. Howes, Jacqueline, 1943-
III. Title.
TA660.C3C45 1984 624.1'774 83-10077
ISBN 0-86095-040-9

Printed in Singapore by
Tien Mah Litho Printing Co (Pte) Ltd

# Contents

# Preface

This book is the result of a study undertaken in 1981 as a prelude to the preparation of an architectural design scheme involving the use of a series of cable net structures. During this study we found that there was much published information available which demonstrated the geometric form and overall architectural implications of net structures; also accessible were a number of detailed mathematical treatments of the analysis of stresses in cable systems.

However, we felt keenly the lack of a simple introductory treatment of the components employed in these structures. Such a work, we believe, would go some way towards facilitating a wider understanding of cable nets. It will also, we hope, encourage those who are interested in the form but who are hampered by a sense of unfamiliarity, to proceed with investigations and proposals with a little more knowledge and confidence.

# Acknowledgements

Figs. 183–6 are based on illustrations from Conrad Roland (1970) *Frei Otto: Structures*, Longman, London. Figs. 187 and 189 are from I.L.8 *Nets in Nature and Technics*, The Institute of Lightweight Structures, Stuttgart.

*Chapter 1*

# Introduction

## Scope

It is intended that this book will provide an introductory overview of the technology of modern lightweight cable net structures, by means of brief descriptions and illustrations of the major components of such structural systems. After a short historical background, which introduces the use of cable nets, a classification of tensile structural systems is given. This, together with the definition of terms section, gives an understanding of specialist language as it is used within this volume.

The performance of each major element is described and the most important methods of construction illustrated. This is accompanied by text which describes appropriate conditions of use for the elements so that the application of each type is understood. Included are sections on the protection and the costs of each element.

A more detailed engineering analysis is not within the remit of this volume, nor is the study of the constraints on built form which the adoption of such a technology implies. It is hoped the reader may take up such studies through reference to some of the structures briefly illustrated in the appendix, and through further reading.

## Development of net structures

Nets as tensile-stressed structures consisting of cables and nodes have been made by man for different purposes since the invention of knotting and weaving. Initially uses were limited to non-structural ones, e.g. fishing and hunting nets. References to Viking vessels include one of the

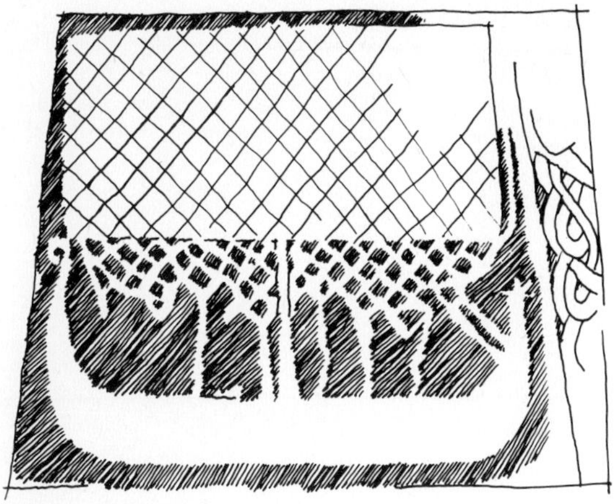

Net supporting the sails of a Viking vessel. Taken from picture stones from southern Sweden

Fig. 1

Roman army tents with net reinforcements. Taken from a relief on the Trajan Column, Rome 113 B.C.

Fig. 2

Bridge across the Niagara 1851–55

Fig. 3

first documented uses of a net as a support, using mesh as strengthening for the sails (Fig. 1). Also, various cultures developed tents, these being the first examples of tension-loaded surface structures. Some of these structures had cables sewn in as reinforcing to the membrane material (Fig. 2). Suspension bridges show the early use of natural fibre cables in tension, supporting footpaths. In the earliest of these the path is laid on the cable and thus follows the curve of the cable. Later, the path was hung from the main cables and was maintained at a constant level by varying the length of a second set of vertical suspension cables. With the development of steel cables, which did not have the problem of the limited life of natural fibre cables, suspended bridges became more widespread (Fig. 3).

After a period of refinement of the technology of these suspension bridges the next step was the suspension of roofs. In 1954 Frei Otto published his dissertation *Das Hangende Dach* (The Suspended Roof); this gave a comprehensive survey of all known designs of hanging roofs and of surface structures loaded in tension. In turn this publication gave an impetus to the further practical investigation of suspended roofs. Architects and engineers throughout the world went on to carry out theoretical and practical work on suspended tensile structures and produced some notable buildings. Some of the most important being the Yale Hockey Rink, New Haven; Dulles Airport, Washington DC; and the French and US pavilions at the Brussels World Fair and the Skidmore, Owing and Merrill building with a

suspended roof for the Baxter Laboratory in Chicago, Illinois (Fig. 4).

In 1955 Otto erected a cable and membrane pavilion for the Federal Garden Exhibition at Kassel (Fig. 5). Several small pavilions and exhibition tents followed during the years 1955 to 1960. Otto's projects became larger and more complex during the 1960s, covering larger plan areas and having wider spans using cable nets and stressed membranes. This work led to the German Pavilion at Expo '67 in Montreal (Fig. 6). Frei Otto's work since has included, in 1968, the building now used as his studio which was an experimental cable net structure with a permanent cladding of wood planking, insulating boards and asbestos cement tiles for the Institute of Light Surface Structures.

In 1969 the work was started for the main Olympic Stadium, the sports stadium, the Olympic pool and the connecting walks for the 1972 Olympic Games complex in Munich.

The lead provided by Frei Otto has given other architects and engineers the incentive to use lightweight tensile structures.

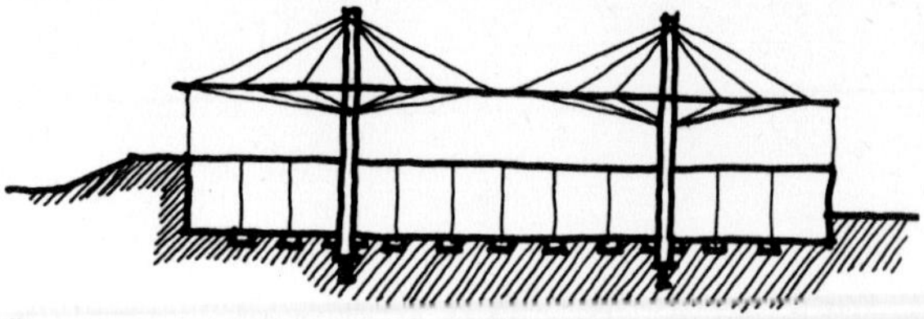

Baxter laboratories,
Chicago, Illinois

Fig. 4

Bandstand – Federal Garden
Exhibition, Kassel

Fig. 5

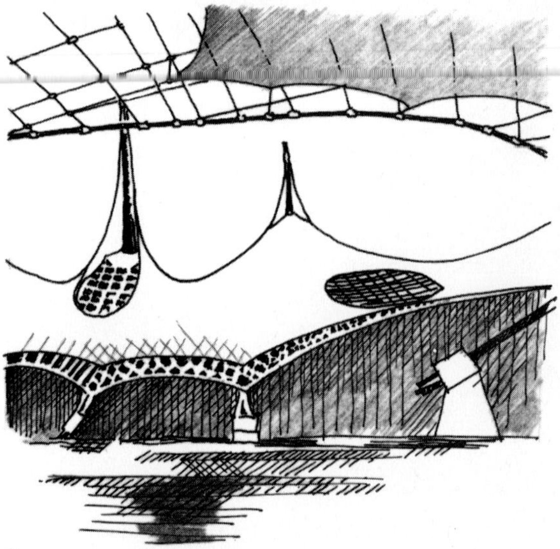

German Pavilion at the World
Expo' 67, Montreal, Canada.

Fig. 6

## Classification of structural systems

Structural systems have been classified according to the following criteria:

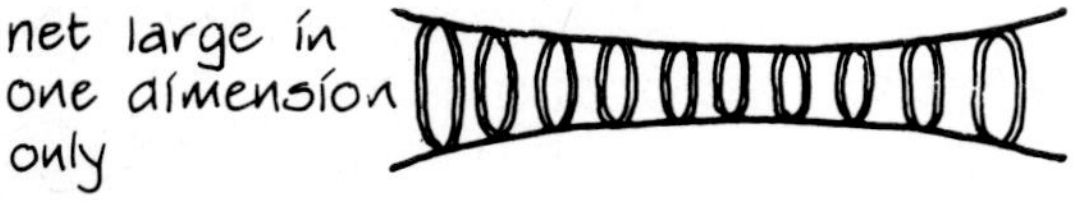

Fig. 7

### *Dimensionally*

*One-dimensional* – large in one dimension only; for instance, cables, columns and beams (Fig. 7).
*Two-dimensional* – equally large in two dimensions; 'surface' structures, plates, nets and membranes (Fig. 8).
*Three-dimensional* – equally large in three dimensions; space frames, spatial cable nets (Fig. 9).

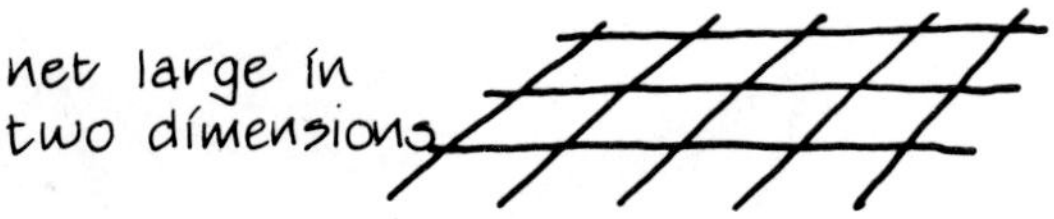

Fig. 8

### *By the loading*

By tensile or compressive force or bending moments. Otto also classifies the forces of magnetism, attraction, rotation or repulsion.

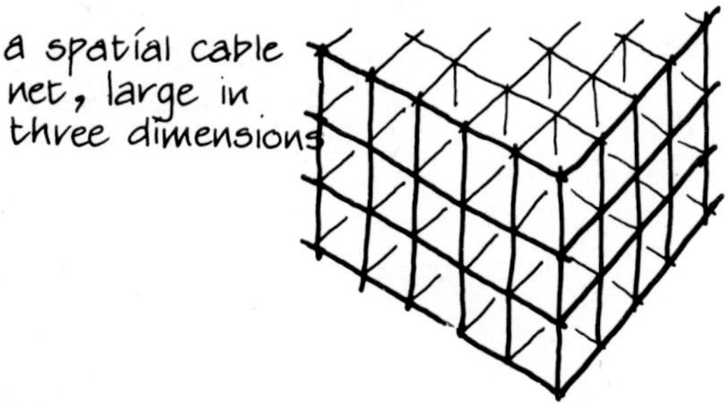

Fig. 9

### *According to condition*

By material characteristics, solid, plastic.

In this volume consideration is given only to the two-dimensional structure in tension. Within this class there are two categories of cable network.

### *1. Continuous membranes*

In this category the cladding of the net is taken into account as contributing structural stiffness when the net is loaded. Thus the cladding and the net are treated as a continuous whole.

### *2. Discrete systems*

Here the cladding is disregarded when calculating the reaction to loading. All analytical procedures are on the basis that the cable network is a discrete system.

**The elements of a cable net structure are supports, cables and anchorages. Connections and coverings are fundamental to the total structure.**

*Chapter 2*

# Cables

## Analysis

A cable maintains its stable form only under tension. If a cable is prestressed then it will take its typical minimum position under load. The cable must be able to take the dead and imposed loads of snow, wind and rain without undergoing undue movement. The amount it is prestressed determines the size of the cable and the size of the anchorages and supports.

A cable transmits loads by developing direct tension. At any point on a cable hanging under gravity the product of the horizontal component of the cable tension and the vertical distance from that point to the cable equals the bending moment which would occur at that section if the gravity load was acting on a beam of the same horizontal span as that of the cable.

## Suitability of use

When deciding on the type and size of cable to use the following criteria and properties have to be taken into consideration:

- level of prestress needed in the cable
- short, medium and long-term loads
- permanent stress
- tensile strength
- permanent vibration strength
- knot strength
- tearing strength
- behaviour in wet and dry conditions
- behaviour in heat and cold
- water absorption
- elasticity
- coefficient of expansion

The performance which is required of the cable must be specified and this will define the standards of the above.

Manufacturers generally provide information concerning the strengths and characteristics of their individual products.

## Cable types

Metal fibres are spun into a yarn or ply, which in turn is combined into strands. A number of strands are then combined in a variety of ways in order to increase the ability of the cable to transmit tensile forces, thus forming stranded, braided or bundled cables (Fig. 10).

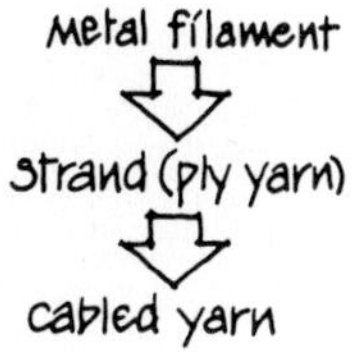


Fig. 10

### *Bundled cables*

Bundled cables are very strong, as the bearing capacity is equal to the sum of the bearing capacity of the individual wires (Fig. 11). These tend to be used for heavy-duty work, such as edge cables and guy cables, where a small elongation is necessary.

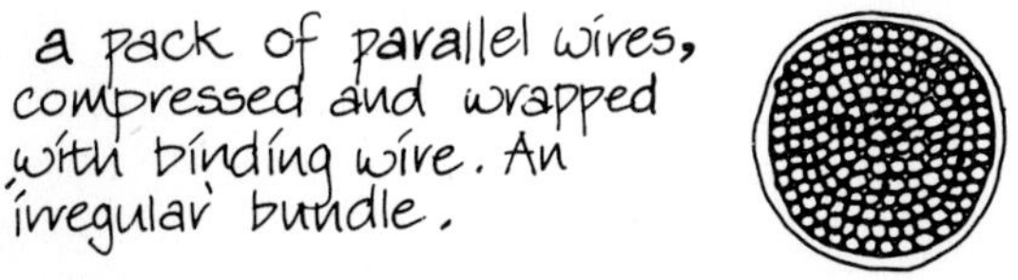


Fig. 11

The use of a metal core gives a much stronger cable but increases its weight. Flexibility depends upon the number and diameter of the single wires. A cable with a small number of thick wires (Fig. 12) is less flexible than a cable with a greater number of thin wires (Fig. 13), and therefore requires a larger radius of curvature. However, it is extremely resistant to wear and corrosion. As the number of wires increases so does the flexibility, but at the same time the strength decreases.

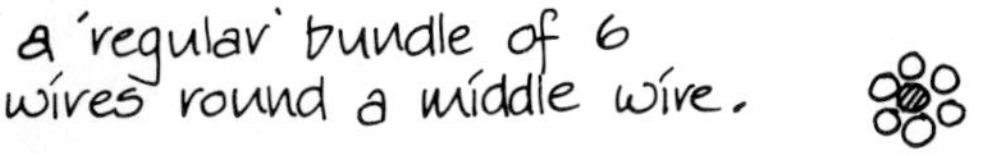


Fig. 12

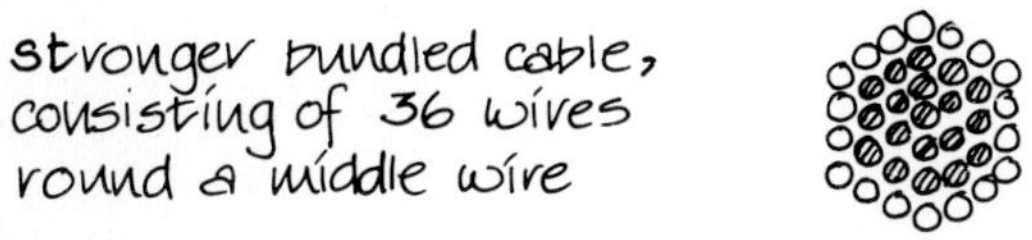


Fig. 13

### *Braided cables*

Braided cables (Fig. 14) (plaited) have a slightly greater tearing length (where the tearing length is defined as the length of cable which will hang under its own weight before breaking) and smaller elasticity than stranded and bundled cables. However, the main advantage of a braided cable is that it has only a small loss in strength when knotted. This makes it suitable for such jobs as mooring cables.

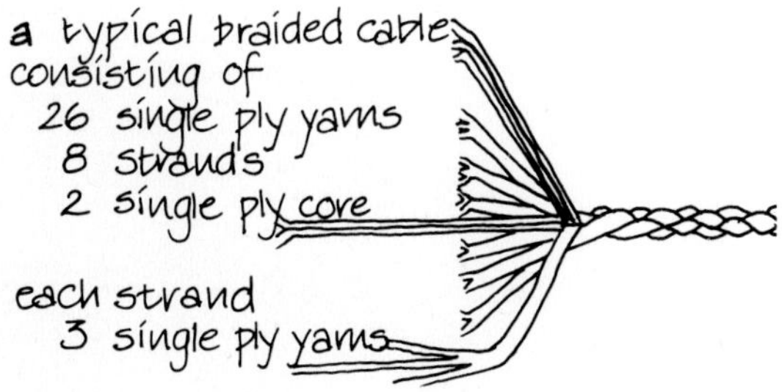


Fig. 14

*Stranded cables*

The properties of stranded cables can be altered by changing the strand size (Fig. 15). Very small strands give a higher elongation and small modulus of elasticity. By increasing the strand size the elongation becomes less and the modulus of elasticity becomes greater.

The standard stranded cable has several wires stranded around a core wire and has a very small elongation under high loads (Fig. 16). They are suitable for use as guy ropes and tie ropes.

Round stranded cables, have stranded cables wrapped round a central core of fibre or metal (Fig. 17).

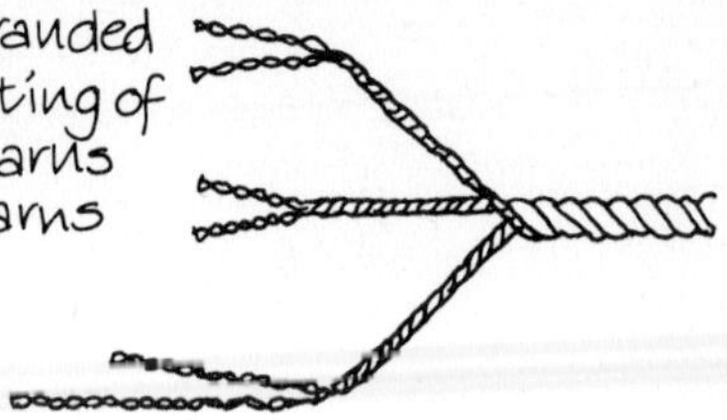


Fig. 15

Fig. 16

Fig. 17

## Non-structural considerations

### *Protection of cables*

Almost all materials which can be used in net structures are subject to corrosion, and/or other forms of deterioration, so they have to be protected from the atmosphere.

*Sheathing*
Sheathing can be taken to include all types of protection which form an outer shell to the cable; for example, plastic sheathing and 'high body' zinc-based paints (Fig. 18). Water vapour can penetrate through joints in the sheathing material and, due to flexing of the cables, through cracks in paint films. This leads to a regular maintenance problem.

*Impregnation*
Plastic resin impregnations are used in order to penetrate all round the individual strands of the cable (Fig. 19 (a)), and because of this property they are particularly suitable for the protection of joints. Care has to be taken in order to ensure that no air pockets are formed, and greater protection is afforded if the material is extended to form a cover to the exterior of the cable (Fig. 19 (b)).

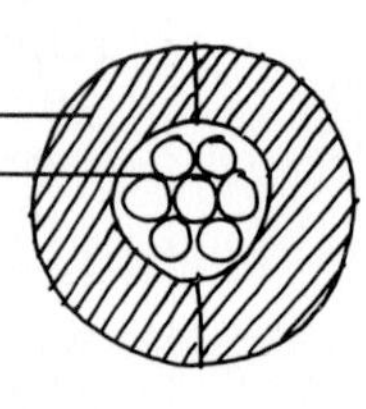


Fig. 18

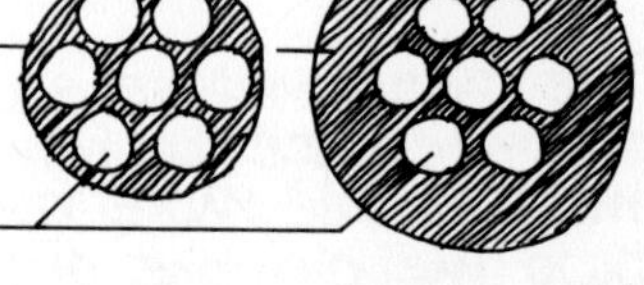


Fig. 19

THE LIBRARY
GUILDFORD COLLEGE
of Further and Higher Educatio

*Electrolytic potential*
Metal cables can be protected from the worst effects of corrosion by the application of an electrolytic potential along their length. The cable has to be insulated at supports and anchorages in order to prevent current drain, and this can prove difficult to arrange in a highly stressed structure.

*Positioning of cables*
Cladding placed over the cable mesh protects the structure from the external atmosphere, but exposes the cables to the corrosive effects of condensation and internally generated pollutants (Fig. 20).

Alternatively, cladding can be placed on the interior of the mesh so the roofing is suspended below the cables, in which case the cables must have greater protection, or a higher maintenance level must be accepted (Fig. 21).

*Stress corrosion*
Steel cables under stress are more likely to suffer corrosion at an increased rate. Since one of the objectives of using a cable net is to economise on the quantity of structural material used, high stress levels are commonly encountered. Therefore the stress corrosion phenomenon must be borne in mind when specifying protection levels.

*The electrochemical series*
Careful consideration must be given to the prevention of electrochemical corrosion where more than one metal is used in the total structural and cladding system (Fig. 22). Where metal-to-metal contact occurs and if the metals are of different types, the base metal will rapidly corrode. For example, aluminium, a common building material, will decompose in preference to most other metals and in particular should never be used in contact with copper or its alloys. The insertion of an electrically inert material between the two metals is the only sure way of avoiding the problem.

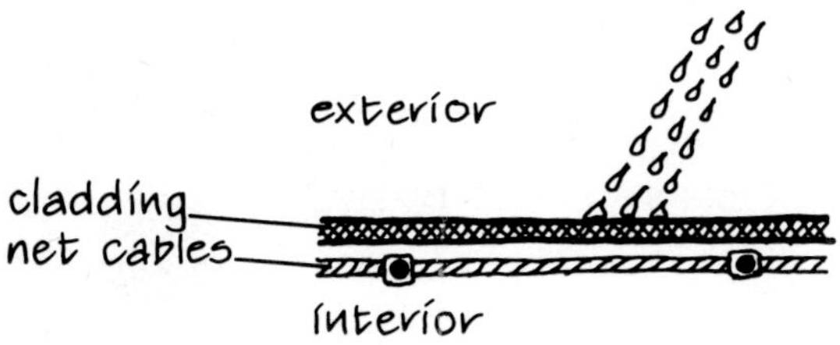


Fig. 20

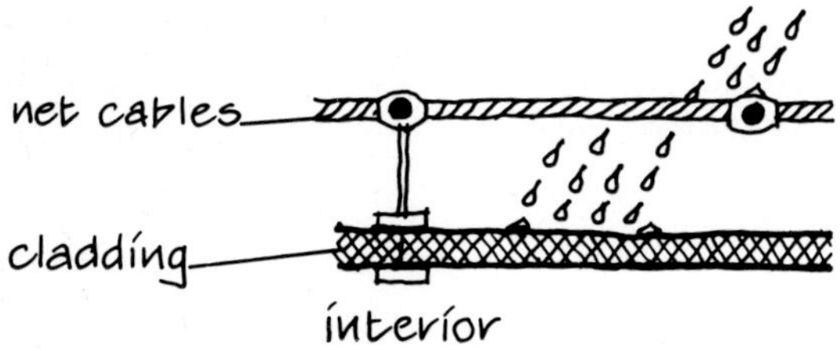


Fig. 21

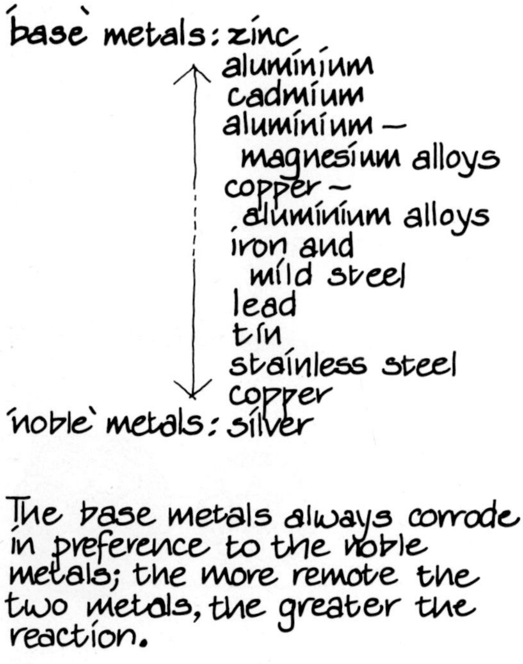


Fig. 22

### *Cost implications*

The total cost of the construction depends on several factors, one of which is the cost of the material of the cable. A survey of prices of materials shows that the price of the amount of material needed to transmit 1 Newton over 1 metre multiplied by a safety factor of 2 (for steel) and 2.5 (other fibres) shows steel to be the most economical and polypropylene rope to be the cheapest.

It can be envisaged that in the future increasing use will be made of high tensile glass or carbon fibres in lightweight cable construction, as has already occurred in membrane structures. This could reduce the cost of cable material and more particularly the cost of cable protection throughout the life of the building. However, the production of these materials in structural cable form is still experimental.

*Chapter 3*

# Connections

## Knots

A knot is the basic building component which forms the main node element of the net. It is of great importance to the load-carrying capacity and to the behaviour of the net under load. The design of the knot is dependent upon:
(a) cable material;
(b) cable size;
(c) the stress in the cables which must be transmitted through the knot.

The knot acts by transferring within it the forces in the cable. The amount of force, known as the sliding force, which it can absorb is dependent on the clamping force which the knot exerts on the cables and also on the coefficient of friction between the knot and the cable. The clamping force itself is dependent upon the amount of transverse pressure the cable can bear. The value of the coefficient of friction can be improved by making the surface areas in contact rougher, or by pushing the cable into the knot material, if it is softer than the material of which the cable is composed.

The sliding force of the knot must be sufficient to keep its position during assembly and at all states of loading once the net is erected. However, when designing for the tearing force it should be ensured that, once a limiting force is reached, the knot will displace. This protection mechanism will prevent the cables from tearing. This in turn can prevent the occurrence of local failure of the structure, as any greatly excessive load will be absorbed by a deformation of a larger region of the net. This will of course cause deformation of any cladding attached to that part of the structure and this must be taken into account.

Various knots are illustrated in the following section with accompanying explanatory notes on manufacture and use. The illustrations cover a cross-section of the most important methods and designs available, starting with simple small-scale applications and progressing to more complex elements. Many have been designed for specific projects. The performance of an element within a net structure should be considered as an individual case. The illustrations are given to provide knowledge of basic techniques and when these are suitable for use.

In large structures sliding clamps can be added in specific areas. These have a defined sliding force and can be used to prevent overloading of individual cables.

Fig. 23

### *Simple knotless links*

*Suspension links*
Suspension links are manufactured mechanically and are therefore relatively cheap to produce; however, they only remain stable under stress. Their main use is in retaining nets and catch nets (Fig. 23).

Fig. 24

*Interlacing links*
Interlacing links are more stable than the simple suspension link but they too still only keep their shape under stress. Again these are used in small lightweight nets (Fig. 24).

### *Knotted connections*

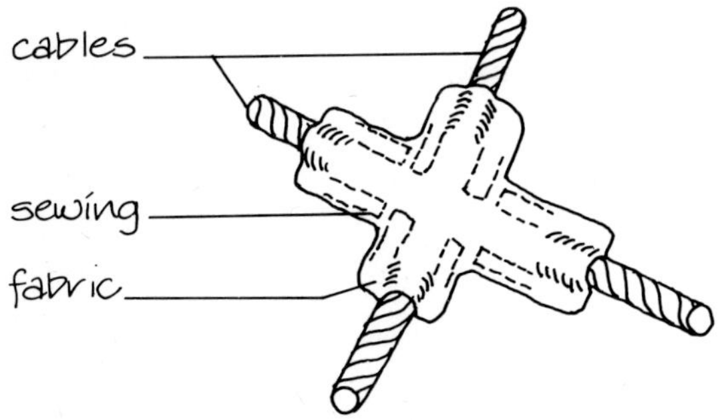


Fig. 25

*Sewed connections*
In a sewed connection the cables are sewn into strips of fabric. If a non-rotating knot is required the fabric would then be sewn in the manner shown in Fig. 25, whereas limited rotation can be provided for by riveting through the fabric. This form of knot is used for reinforcing stressed membranes and small-scale stressed cable nets, e.g. exhibition tents (see Appendix – Bandstand at Kassel).

*Glued connections*

Glued connections can be used in areas of low stress. The knot is composed of a number of elements which, when fitted together, form a cable-enclosing unit. The individual elements can then be glued using epoxy resins or similar high performance glues (Fig. 26).

*Welded connections*

Figure 27 shows a simple welded connection between two cables which can only transfer small loads. Figure 28 shows a spot-welded clamp made from sheet metal. Because of the greater weld area and greater bearing surface on each cable it can obviously be used in higher stress situations than the simple weld connection. This is a highly economical system but there can be difficulties due to the cable material becoming heated and thus having its tensile strength impaired. If the cables are made from high-tensile steel they must have profiles of weldable material.

*Casings*

Casings are non-rotational fittings and must transfer the loads in the knot to the cable by friction alone (Fig. 29).

*Cast knots*

Cast knots are formed by casting low-melting-point alloys or high-tensile plastics around the cables (Fig. 30). To protect the knot from displacement an interior reinforcing strand must be added. This is a time-consuming method but it is economical on material. The disadvantage is that this can only form a non-rotating knot and as such is only suitable on small stress nets (Fig. 31).

Fig. 26

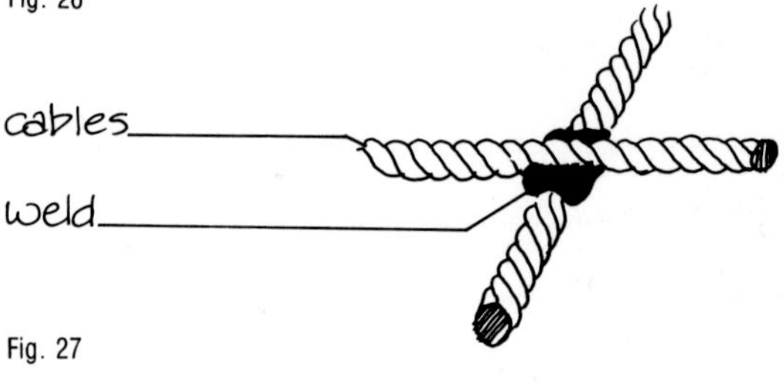


Fig. 27

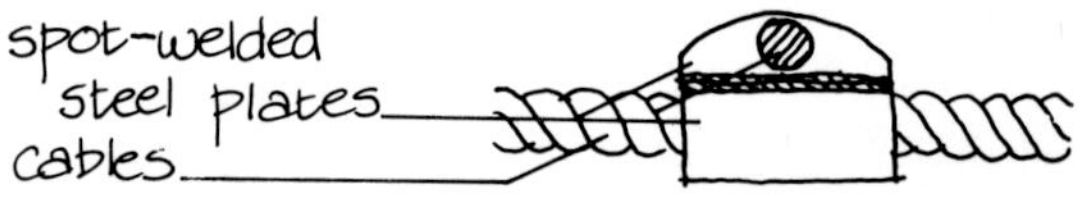


Fig. 28

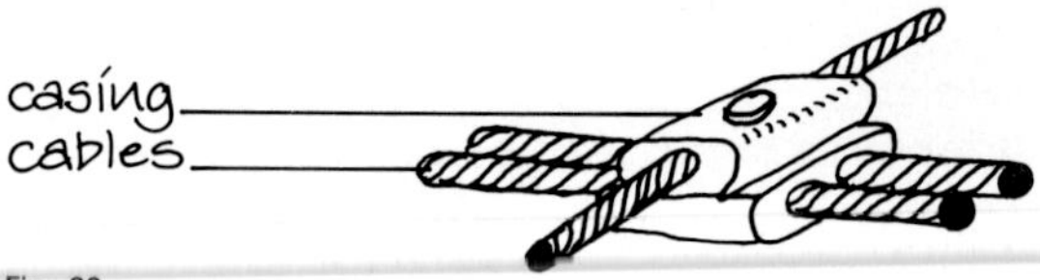


Fig. 29

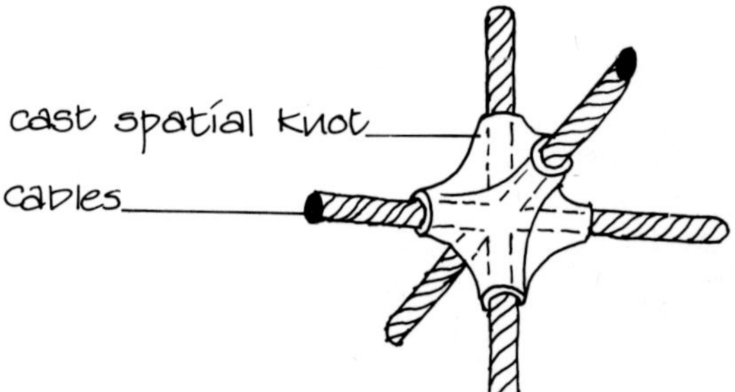


Fig. 30

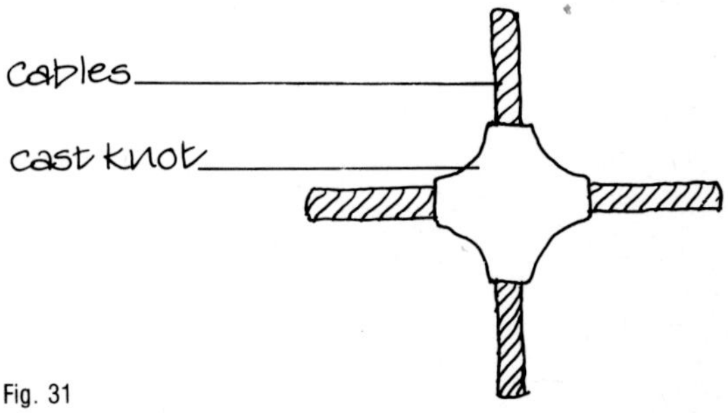


Fig. 31

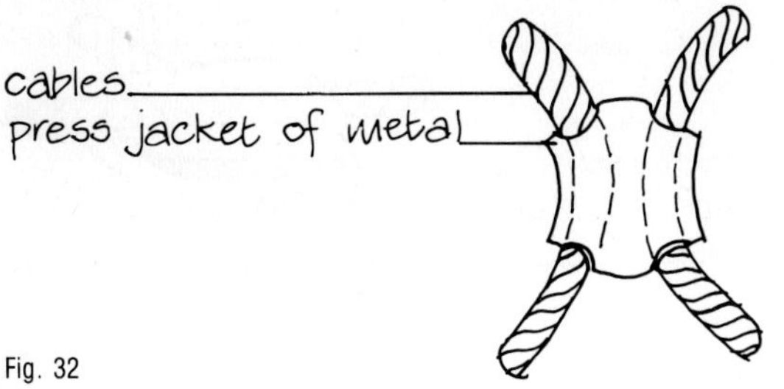


Fig. 32

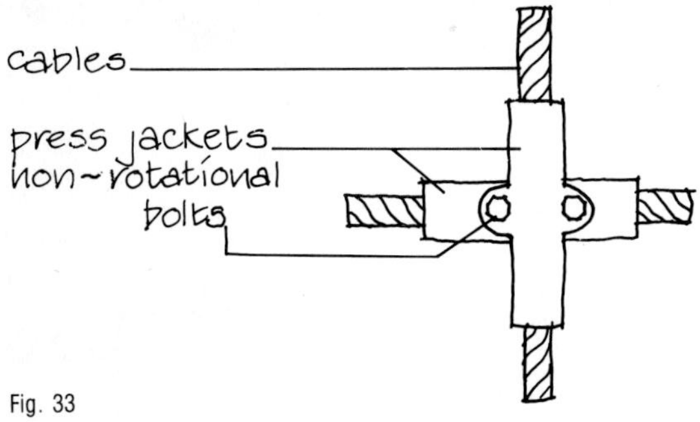


Fig. 33

Fig. 34

Fig. 35

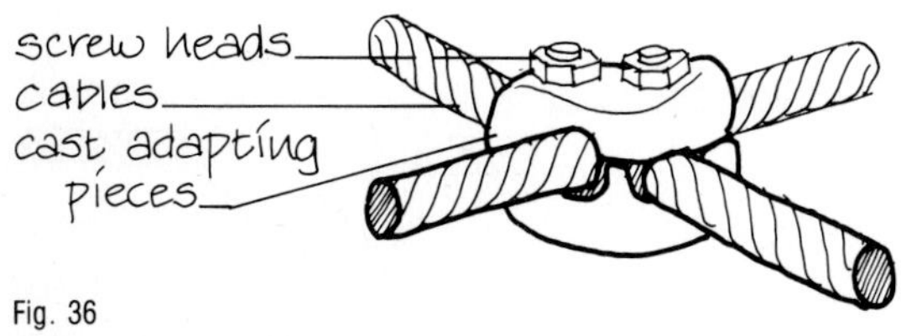


Fig. 36

*Jacket knots*

A press jacket of malleable metal (often aluminium) can be used when cables do not cross; for instance, in extensible nets. The open sides of the jacket are offered up to the cables, closure being effected by the operation of a suitable pressing tool (Fig. 32).

It must be remembered that where aluminium is used mild steel must be sufficiently protected to prevent an electro-potential being set up and corrosion resulting.

A more complex jacket connection uses a press jacket on each cable which is joined either with a rotating eye, or bolted together to form a non-rotating knot (Fig. 33).

*Screw-bolted knots*

The most simple screw-bolted knots are angled bolts with a threaded terminal and an eye for inserting the second bolt. The pressure on the cable is distributed by using an adapting piece between the bolts (Fig. 34). Figure 35 shows screw-bolts used on the Brooklyn Suspension Bridge in 1865.

*Screw-clamps*

Screw-clamps are a more robust development of the screw-bolted knot where the adapting piece is made up of several forged or cast parts. These are then connected by the use of one or more screw-bolts. The transmitted force within the knot can be altered by loosening or tightening the screw (Fig. 36). This is an example of a non-rotating clamp as used at Montreal (see Appendix). The screw heads must be given special protection to avoid water penetration.

The basic design can be varied to cater for rotation, and in this case for double cable nets. The parts are forged. These can be mechanically placed on to one set of cables at the correct intervals with manual completion of final assembly (Fig. 37).

In the variation shown in Fig. 38 the two parts can both be placed on the cables mechanically at the correct intervals. Final assembly is achieved by means of the four screws.

Fig. 37

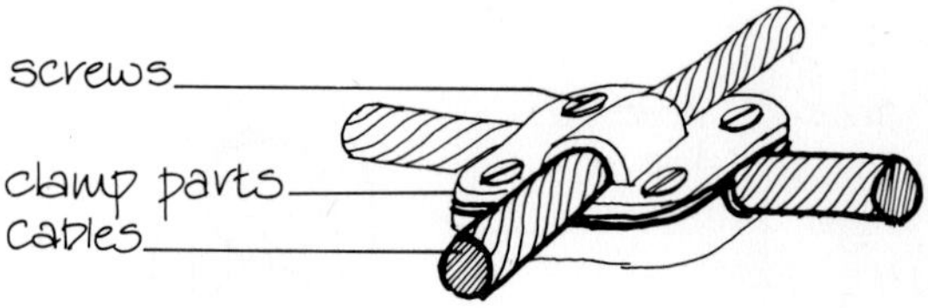

Fig. 38

## Joints

Joints occur where two cables are joined together longitudinally, either to extend the cable length or in order to provide a sliding clamp.

The simplest of joints are for use with highly flexible cables of natural fibres – flax, hemp, jute, manilla – and synthetic polymer fibres such as polyester, polypropylene and polyethylene. Metallic fibres are too stiff for the employment of these forms of joints.

### *Simple link joints*

Figure 39 shows simple interlacing of cables. These joints rely on the tension being maintained for continued structural stability. If the tension is released then the joints untie themselves.

Many other forms of knots and splices can be used to join these cables, but in all cases their use is confined to nets subject only to small loads; for example, retaining nets (Fig. 40).

Fig. 39

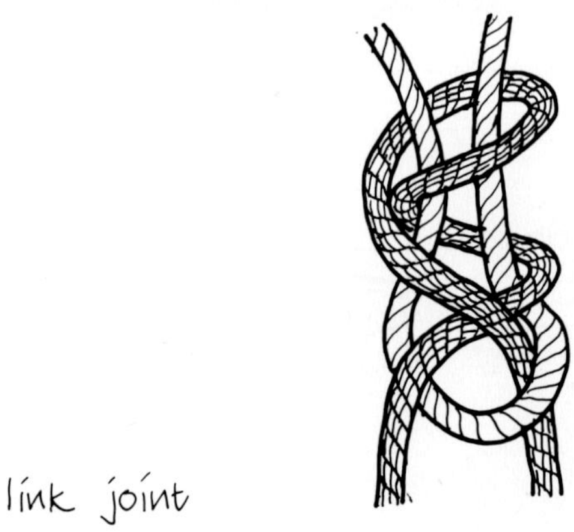

Fig. 40

### *Clamps*

Joint connections used for metallic filament cables such as steel, stainless steel, nickel, copper or molybdenum have to employ clamps of some form.

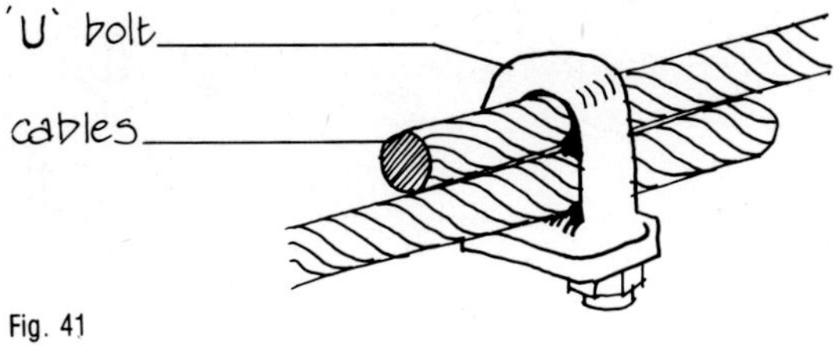

Fig. 41

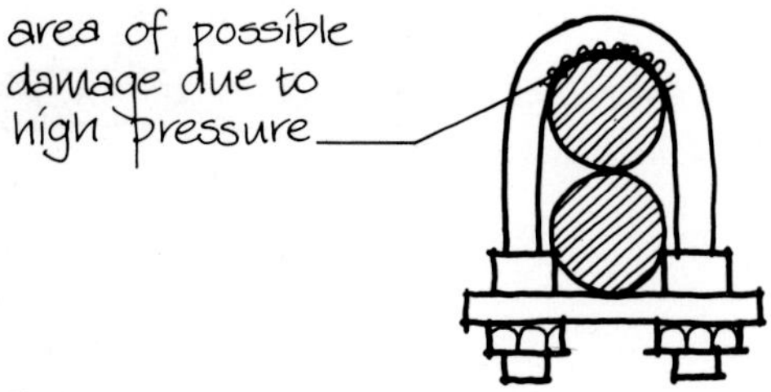

Fig. 42

Fig. 43

Fig. 44

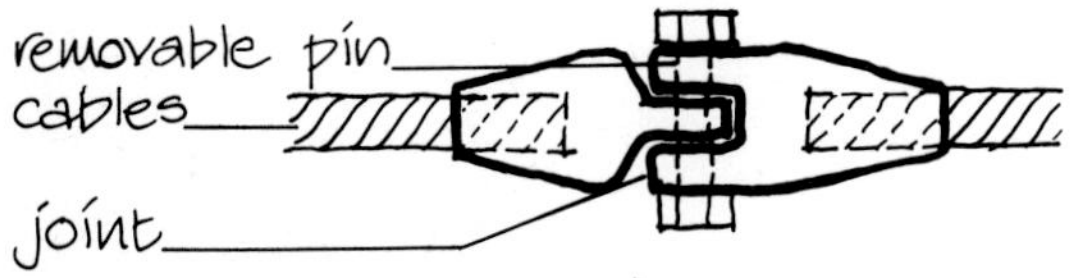

Fig. 45

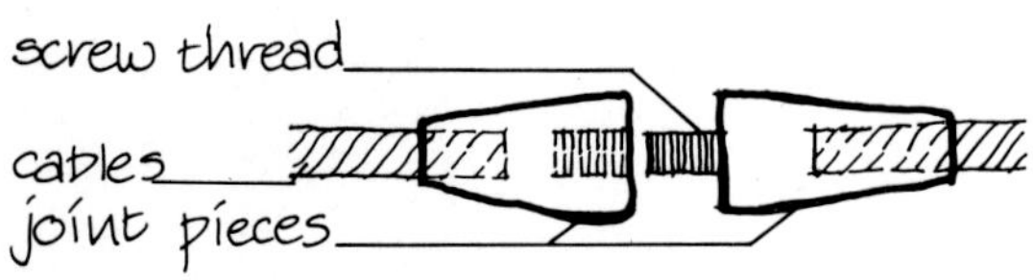

Fig. 46

*Cable clamps*
Cable clamps are 'U'-shaped bolts which hold the cables together (Fig. 41). These must often be used in series in order to transmit the forces. The use of this form of clamp can cause danger of local damage to the cable due to the high compressive pressures set up (Fig. 42).

*Longitudinal clamps*
Longitudinal clamps distribute the clamping pressure over a larger area and therefore can be designed to avoid local cable damage. The clamp is made from punched sheet parts with screw-bolts (Fig. 43).

*Press clamps*
Longitudinal press clamps can be used as sliding clamps (Fig. 44). The strength of these clamps can correspond to the tearing strength of the cable itself.

*Couplings*
Couplings are quickly connected and loosened and are thus ideal for the construction of temporary structures. They can use a removable pin or a screw thread (Fig. 45).

*Tension lock*
A tension lock can form not only a longitudinal joint but can also be used to adjust the length of, and therefore vary the tension in, individual cables (Fig. 46). In large areas of net they can be positioned so that they form an assembly joint for

separate parts of the net. They are suitable for use in nets with high loading conditions. The lock shown in Fig. 47 was used in the construction of the Munich stadium roof structure (see Appendix).

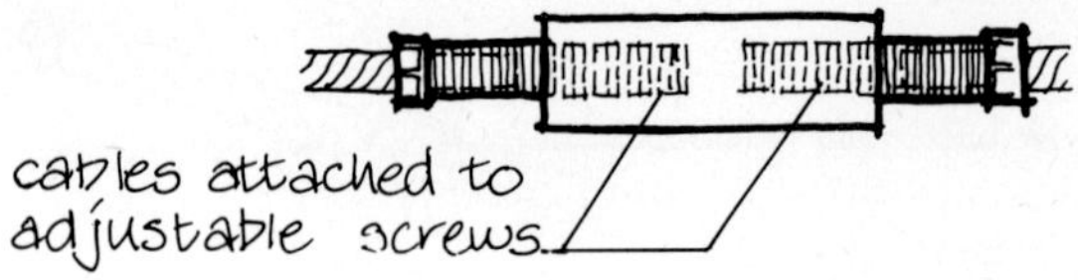

Fig. 47

## Edge knots

The connection between a net cable and the edge cable is an edge knot. The characteristics of an edge knot as opposed to a knot found in the interior of the net are:

- Only one cable runs through the knot, that is the edge cable (Fig. 48).
- The edge cable is usually considerably larger than the net cable (Fig. 49).
- The total stress in the net cable has to be transmitted from the net cable to the edge cable, rather than partially passing through the connection as in an interior knot (Fig. 50).
- The spacing distance of the edge knots on the edge cable depends on the net pattern and is not necessarily regular (Fig. 51).
- The angle of approach of the net cables to the edge cables can vary considerably even in the same local region of a net (Fig. 52).

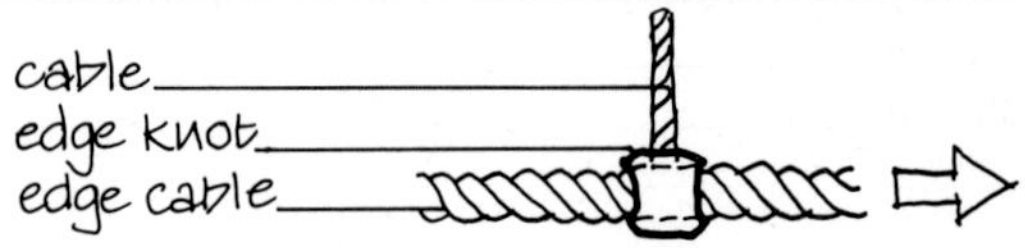

Fig. 48

Fig. 49

Fig. 50

### *Simple edge knots*

Edge connections for nets which are not under heavy strain, such as temporary canopies and retractable roofs, use simple spring grommets and loops. These maintain their position only when tensioned (Fig. 53).

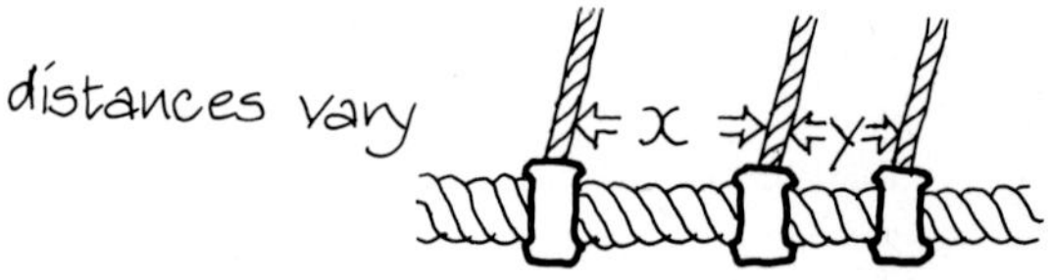

Fig. 51

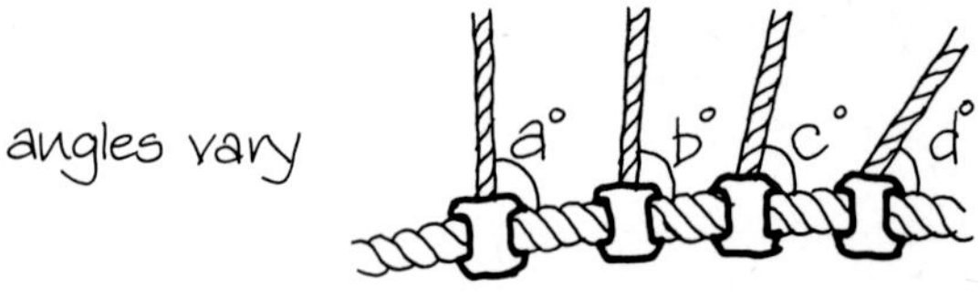

Fig. 52

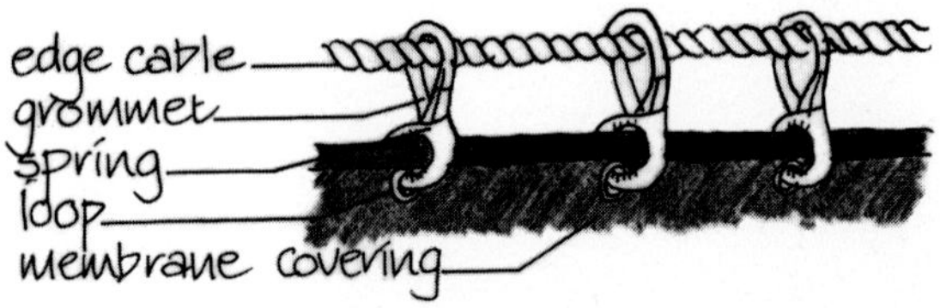

Fig. 53

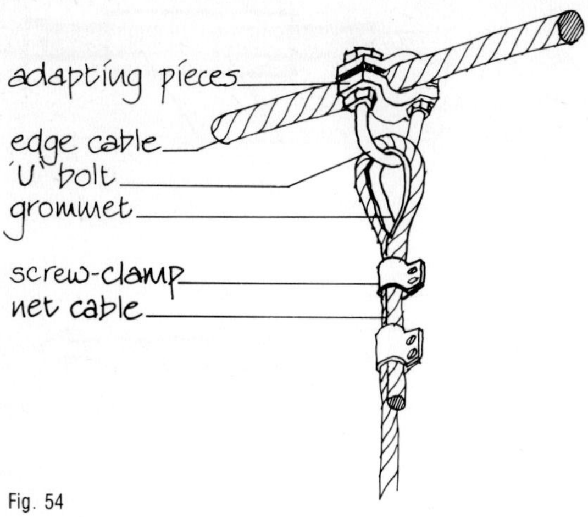


Fig. 54

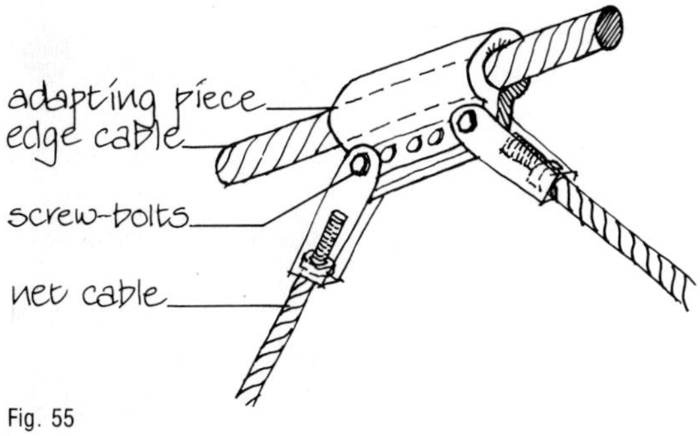


Fig. 55

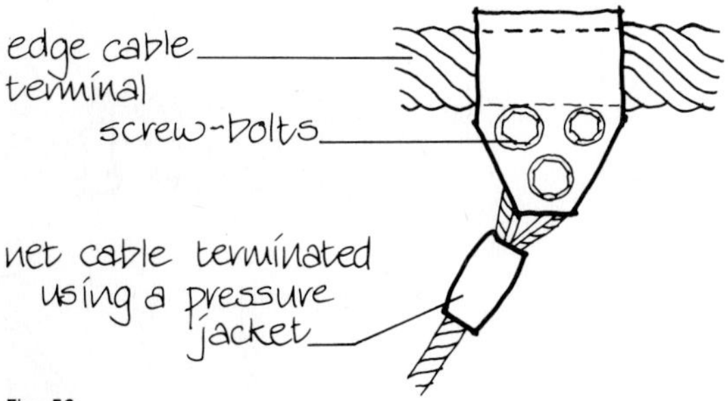


Fig. 56

To maintain the edge knot in position a clamp is formed around the edge cable using a screwed adapting piece. The net cable is fitted over a grommet which pivots on a 'U'-bolt and is fastened back with screw clamps (Fig. 54). There is no scope for removal or readjustment of the cable when a mistake has been made.

For connections under more stress an adapting piece can be screwed onto the cable using a number of bolts or screws. The outer two bolts are used to support two straps over which the net cables are taken (Fig. 55). This system is only suitable for nets with a regular edge distribution, as it is not possible to vary the angle of approach of the net cable.

### *Edge net clamps*

This type of clamp (Fig. 56) can be placed at any point on an edge cable. By loosening the terminal screws the position of the edge clamp can be altered or it can be removed altogether. The friction force between the edge cable and clamp can be increased by tightening the terminal screws.

A double edge cable clamp was designed to give a larger contact surface; it has four screws rather than two so the frictional force is increased. Most high loadbearing structures use edge net clamps (Fig. 57).

## Deviation knots and branching knots

A deviation knot constantly changes the direction of one or more continuous cables along a determined flat or spatial curve.

The smallest permissible radius of curvature of the cables within the knot is the most important condition of formation. The parts of the knot can be cast from steel or aluminium, or made from flat or curved sheet metal. These can then be screwed or welded. The openings within the knot which accept the cables should be cone shaped to cope with any cable deformation due to loading on the structure. As the cables go through the knot they must be fixed in position either by clamps within the knot or by ancillary clamps positioned immediately adjacent to the knot.

*Pincers*
The easiest deviation knot to construct is known as a pincer. The cable direction is altered by bolting two suitably shaped sheet metal flaps together around it. The bolts holding the cables together can act as cable guides and as clamps to hold them in position. Figure 58 shows such a knot where two edge cables are turned into a double ridge cable.

The bolts can become the axes for rotating cable discs which improve the guidance of the cable through the knot. Ancillary clamps have to be used to hold the cable in position. This example turns two edge cables into an anchorage. (Fig. 59 a and b)

In Fig. 58 and Fig. 59 (a), (b) the function of the cables is changed; the example shown in Figs. 60 and 61 illustrates cables whose function is maintained whilst they are deviated.

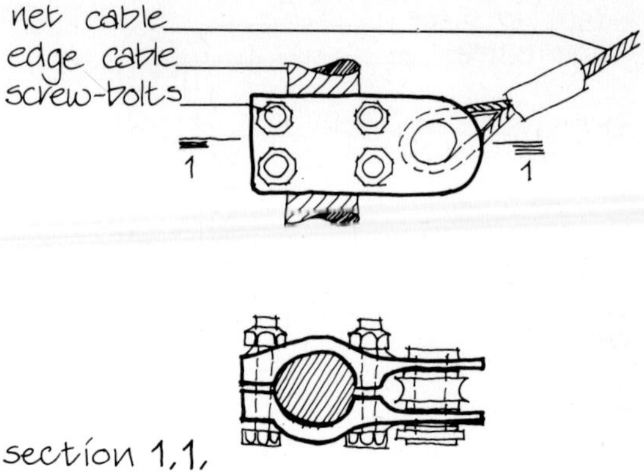


Fig. 57

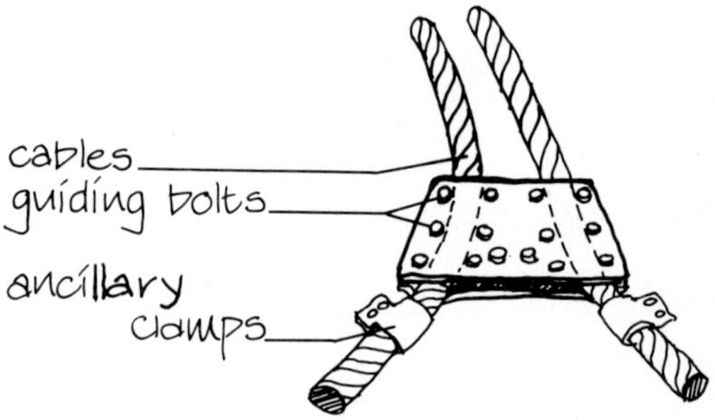


Fig. 58

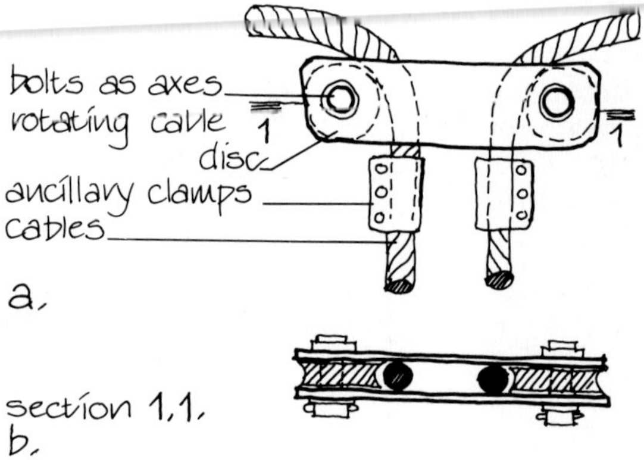


Fig. 59

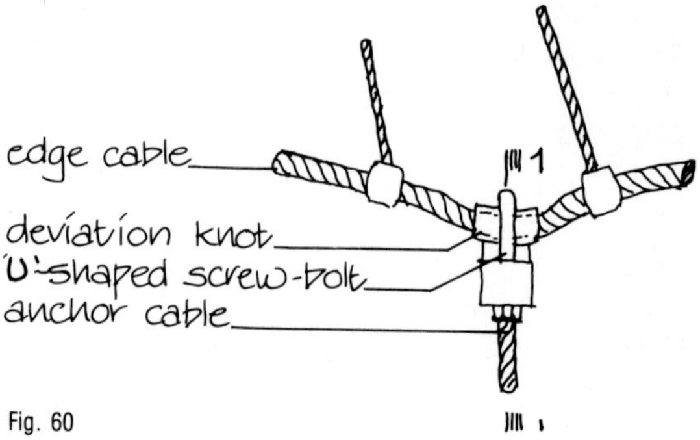


Fig. 60

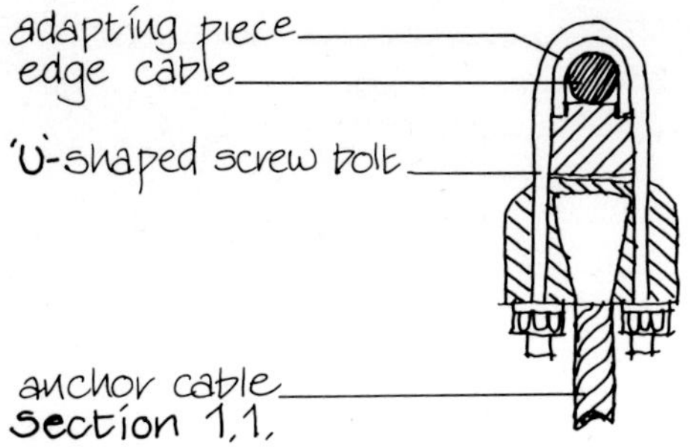


Fig. 61

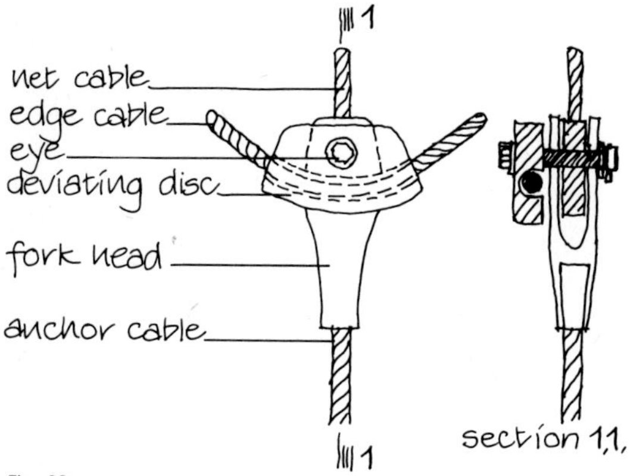


Fig. 62

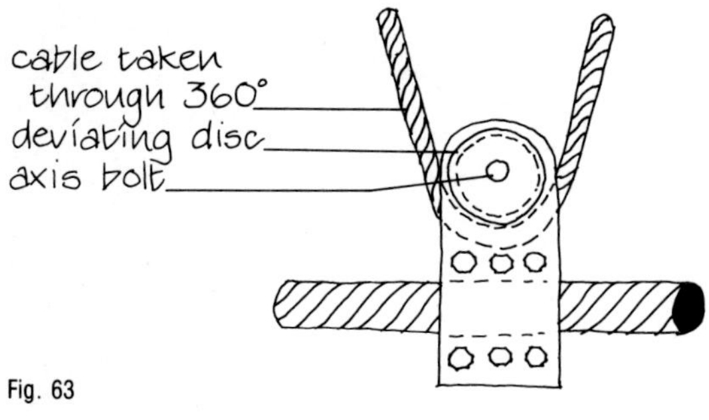


Fig. 63

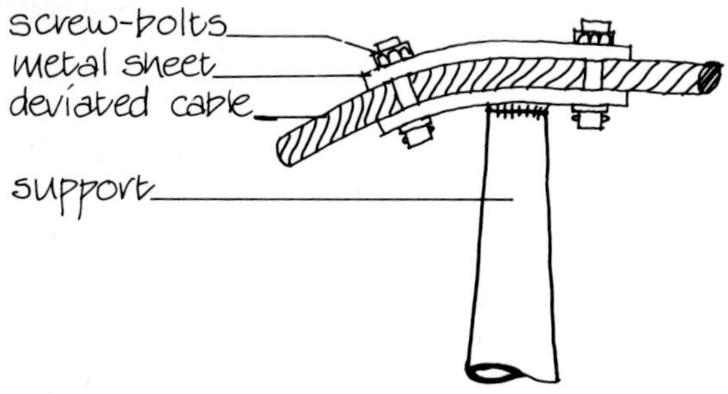


Fig. 64

The edge cable is deviated through a knot which incorporates an anchorage cable. This is fixed to the adapting piece of the knot using a 'U'-shaped screw bolt (Fig. 61).

*Deviating discs*

Deviating discs are used when higher forces have to be transferred within the knot; they also give better guidance to the cable.

A steel deviating disc is joined to the anchoring cable by a fork head and eye. The bolt acts as the rotating axis, which should be as close as possible to the line of the edge cable as this improves the stability of the disc (Fig. 62).

Where a cable is taken round a coil through 360° it is transformed into a double cable within the knot. If the angle of the cables to the anchorage is liable to fluctuation due to load deformation of the net, then the knot must be mounted on a flexible bearing. The pincers and discs are free standing knots as the deviation takes place on the same plane as the net surface (Fig. 63).

Where the deviation does not take place on the same plane and one or more cables are to be deviated onto a spatially different plane then the knot must be supported or suspended.

*Supported deviation knots*

The simplest construction of a deviation saddle consists of two shaped metal sheets connected by screw-bolts. The bolts serve as guides for the cables which are parallel as they enter the saddle and again parallel as they leave the saddle, but on a different plane (Fig. 64).

Figure 65 shows an example of the deviation of an eye edge cable into a double ridge cable on a different plane, which is supported on a mast head.

Where the cables are to be deviated in the centre of a net, for instance, at the top of an eye, the knot can be a cast steel shape as in Figs. 66 and 67 and clamped to the anchoring cable which acts as the support to the saddle.

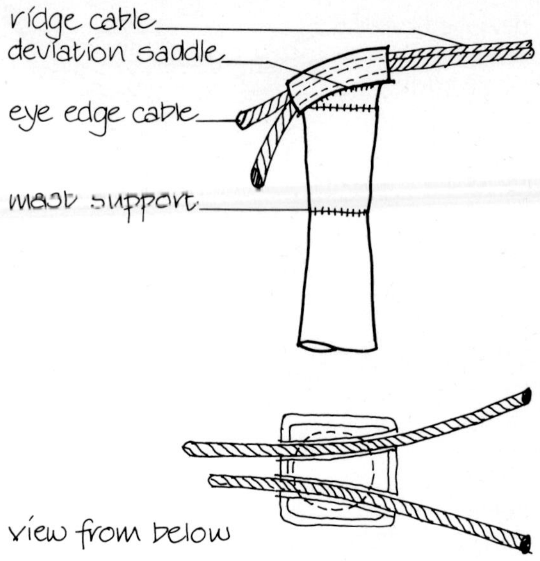

Fig. 65

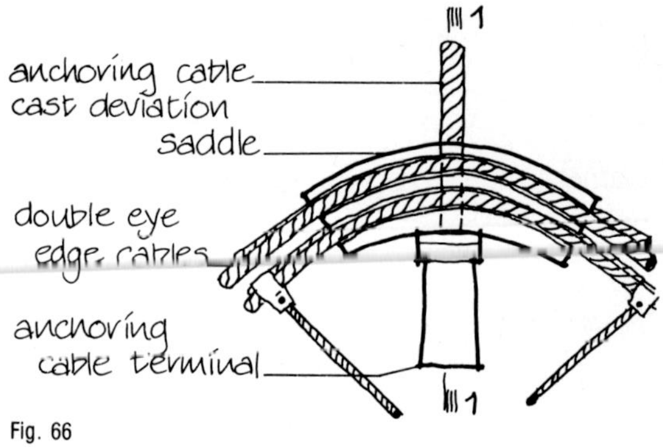

Fig. 66

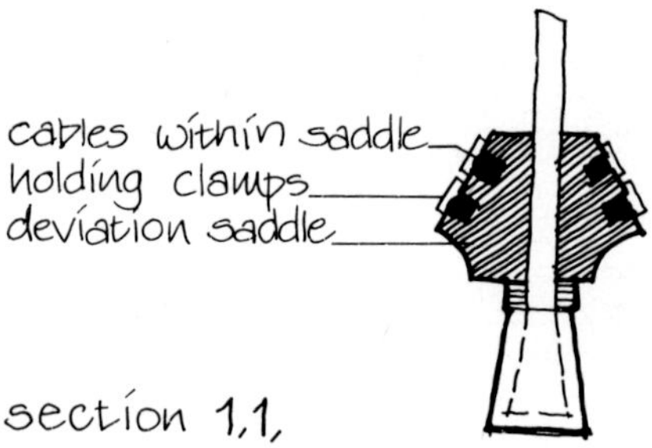

Fig. 67

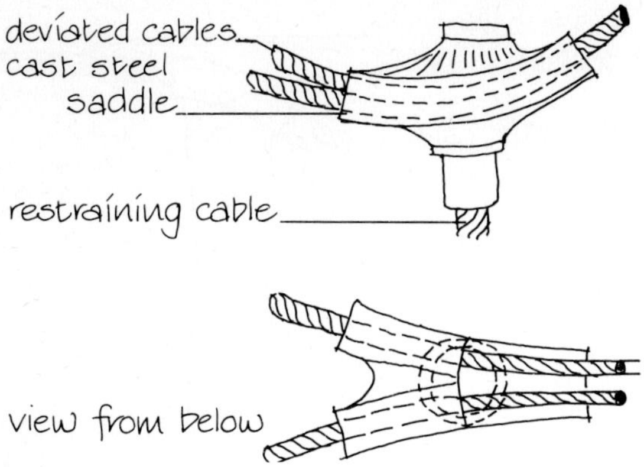

Fig. 68

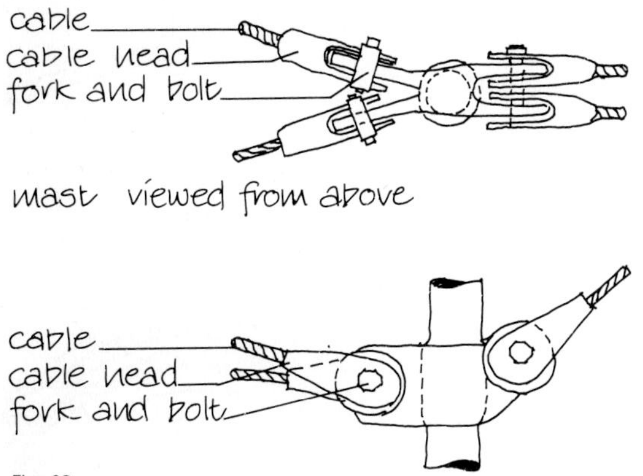

Fig. 69

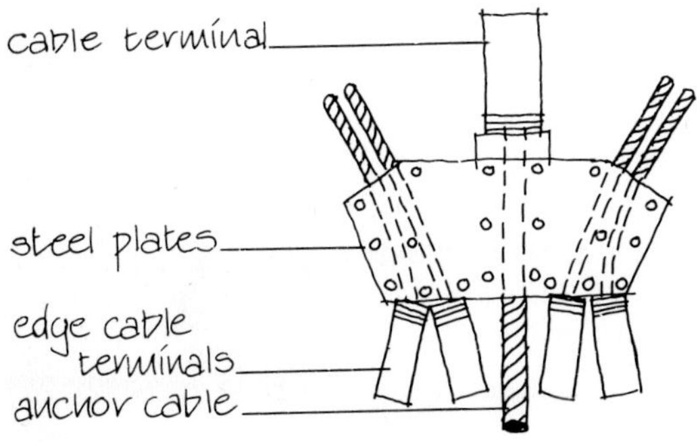

Fig. 70

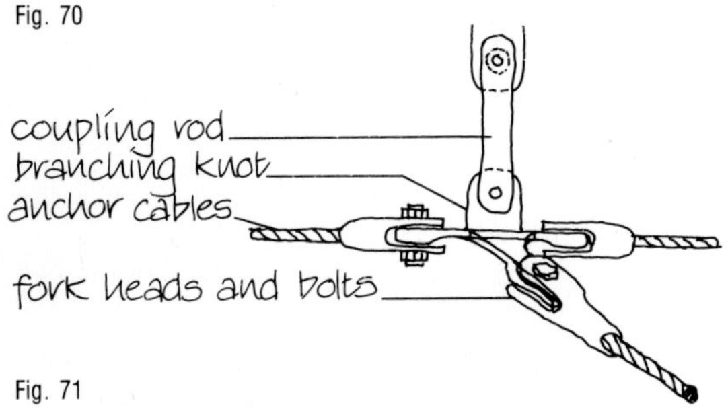

Fig. 71

The cables which are to be deviated at a low point must also be restrained by a support. A low point cast steel deviation saddle is shown in Fig. 68 achieving full spatial deviation of the cables, i.e. the cables alter direction in plan and simultaneously turn into a new plane.

### *Branching knots*

A branching knot has the same function as a deviation knot, but instead of the cable or cables being continuous through the knot, they end within it.

*Low point restraints*
Low point restraints can be cast from steel with the individual cables being held within the branching knot with fork heads and bolts which act as pivots (Fig. 69).

*Acute angle anchorage*
When edge cables approach an anchorage at too sharp an angle to allow deviation to take place, because the cable is not able to accommodate the necessarily small radius of curvature without risk of damage, the acute angle anchorages illustrated in Fig. 70 may be used. The two steel plates are screwed together, trapping the two edge cables, and the anchorage cable. Terminals on all cables prevent pull through under load.

*Multiple cable branching knot*
A knot can accept many cables coming from different directions. In Fig. 71 the knot links three anchoring cables and a coupling rod. The cables are bolted through fork head terminals to the steel plates. It may be necessary to use such a multiple knot for instance in a position in the centre of a net at a low point.

*Combination knots*
Deviation and branching knots can be combined where needed. Figure 72 shows how an edge cable is deviated at the point where three anchoring cables, i.e. the cables alter direction in plan and

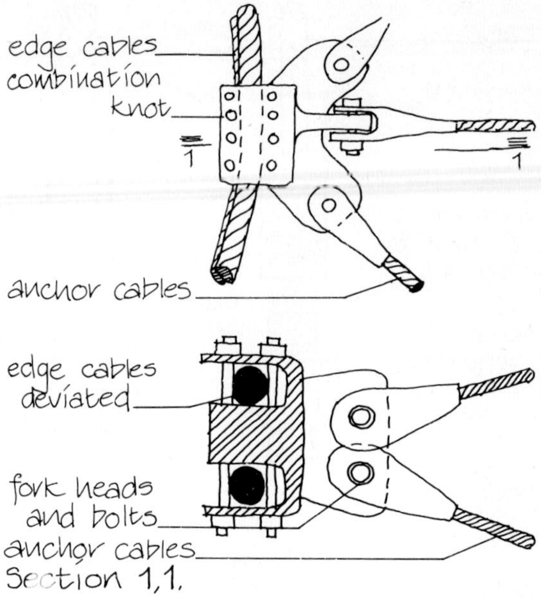


Fig. 72

## Terminals

A terminal is the building component which ends the cable and is used to attach the cable to other building elements.

### *Simple terminals*

For transferring small forces the terminal is usually formed by looping the cable round a grommet, these often being used with natural fibre cables. As the need to transfer larger forces arises then clamped grommets can be used (Fig. 73). Figure 74 shows a terminal using 'U'-shaped screw-bolts to hold the cable. These can cause weaknesses due to local crushing of the cable. Finally, a pressure clamp may be used to give an even force along an extended length of cable.

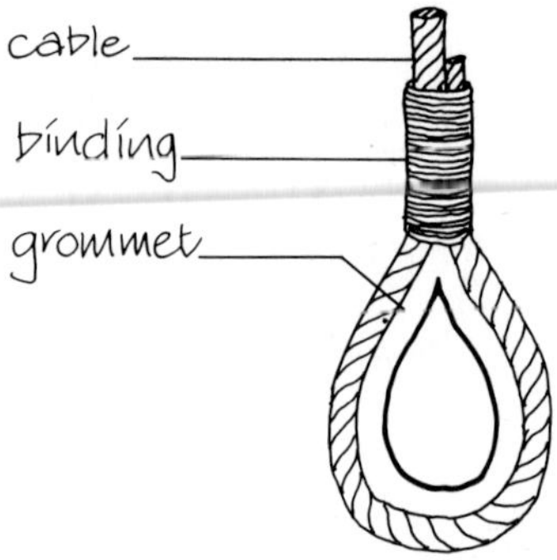


Fig. 73

### *Cable heads*

For the size of forces which have to be transmitted in a large load-bearing structure the terminal is generally manufactured with a cable head. The method by which the cable and head part are joined divides them into two groups: those with pressure jackets and those with grouting jackets.

The type of terminal used depends on the type and material of the cables. At all times the load-carrying capacity of the cable head when subject to vibrational loads has to be taken into account.

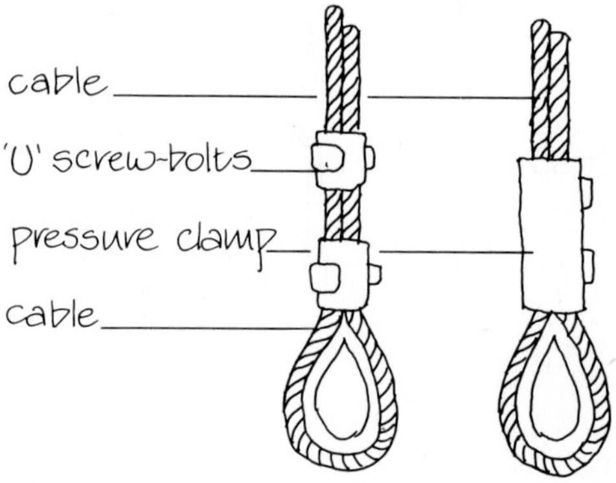


Fig. 74

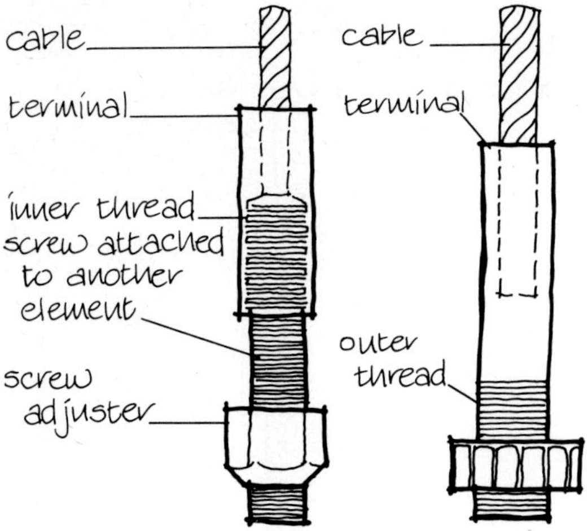


Fig. 75

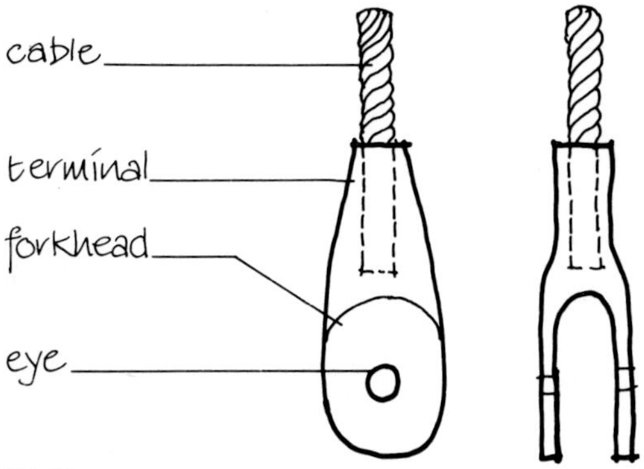


Fig. 76

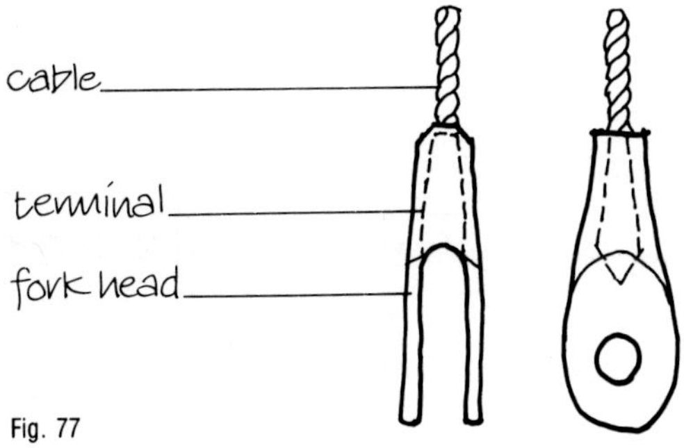


Fig. 77

*Pressure jacket terminals*
For cables of 5 mm–10 mm, pressure jackets are usually used. They can be manufactured from aluminium (if used with steel cables, the steel must be protected from contact with the aluminium) or soft steel, or an inner jacket of harder steel can be inserted. Jackets can be formed with inner threads or with outer threads which can be used to adjust tension once the jacket has been pressed over the cable (Fig. 75).

Pressure jackets with forks can be used as terminals to accept other building elements (Fig. 76).

*Grouting jacket terminals*
For thicker cables, clusters of cables and parallel wires, grouting jackets are used. These are more expensive than pressure jackets, but they give greater stability when transferring large forces.

The end of the cable is formed as a fan-shaped tail which is grouted into a cylindrical jacket. Warm groutings use molten soft metals or alloys of soft metals. One of the earliest materials used for this purpose was lead, due to its low melting point. The grout often has a harder granular metallic or mineral addition which increases its bonding strength. Cold groutings are either cement mortars or synthetic resins, which harden slowly without a noticeable change in temperature; again, an addition of metallic or mineral grains improves the bond between the cables and the casing.

The disadvantage of warm grouting is that it adversely affects the cable material by heating it to a high temperature. The grouting also shrinks on cooling and can therefore loosen its hold on the cable. The advantage of using a cold synthetic grouting, with or without metallic additions, is that the cable is not affected and the shrinking of the grouting is reduced to a minimum.

A cable head formed as a fork is shown in (Fig. 77). A fork can be used to link the cable to a rigid building element, e.g. an anchorage. If the angle of the cable direction is liable to change due to load deformation, then a bearing joint should

be incorporated between the bolt and the eye sheets of the fork. This will prevent undue wear on the connection (Fig. 78).

A grouted-on cable terminal can be shaped to produce various forms of straps for connections to one or more of the building elements (Fig. 79). For example, a forkhead may be formed to take a double deviating disc, which could be used to connect an anchorage cable to an edge cable. Alternatively, a terminal for joining tension rods to the cable network may be simply constructed using this technique (Fig. 80).

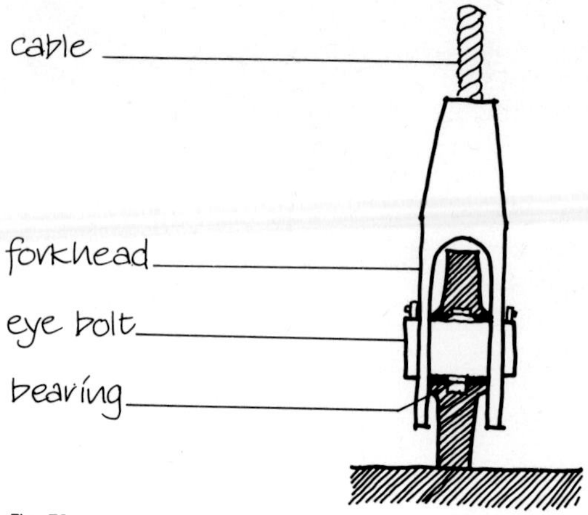

Fig. 78

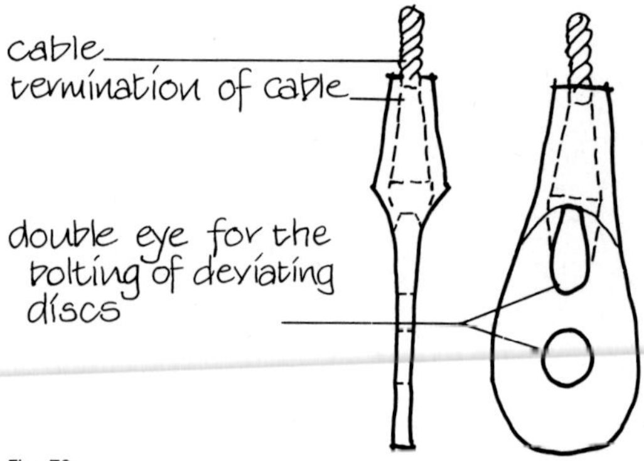

Fig. 79

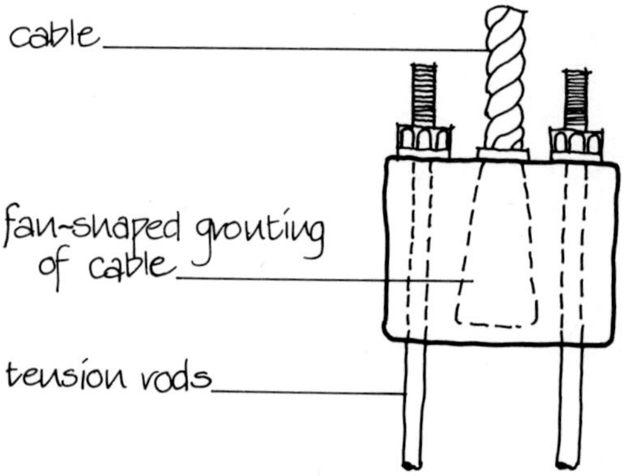

Fig. 80

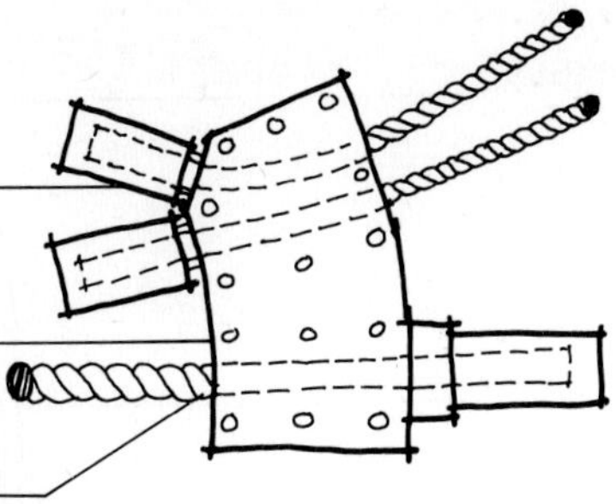

Fig. 81

Finally, simple grouting jacket terminals may be used in positions where there are large pull-through forces; in this case, at the conjunction of anchor and edge cables in a large structure (Fig. 81).

## Non-structural considerations

### *Protection*

The connection is an important element of the overall structure, but because of its form it is prone to corrosion and therefore careful consideration must be given to its protection. The materials used should be checked for electrolytic compatibility. If the knot is made up of parts which are fitted together around the cables then penetration of moisture either as rainwater or condensation is liable to occur. To combat, this, the outer form should be closed and rounded without sharp edges or protrusion at joints.

The position of the net also affects its need for protection. When the connections are on top of the cladding, rainwater from above will be the major problem; if the connections are below the cladding then the possibility of condensation within the individual part will have to be taken into account.

The connections could be positioned within the cladding and this would provide extra protection from corrosion attack, but the problem of access to all the knots for repair or maintenance has to be considered.

### *Cost implications*

Although the cost of each individual connection is small it affects the overall cost of the structure considerably because of the large numbers of units involved.

One factor which clearly affects the number, and therefore total cost, of the connectors is the mesh width. This can be chosen freely, but the implications on the cost of the cables and connections has to be taken into account and balanced against the savings made by the need for fewer cables and connections. As the mesh becomes wider the cable becomes thicker and less flexible, so production, transportation and assembly becomes more difficult and expensive. This also means that higher forces will be transmitted in the connection, so the connection will have to be larger and stronger. This limits the construction possibilities and increases the total weight of the net, which infers a rise in cost to accommodate the need for more material.

Another factor which affects the production cost is the amount of possible prefabrication. As the construction of the connections affects the degree of prefabrication it thus affects the costs. At present the finishing of all structural nets is manual; at most, the knots being placed mechanically on the cables at the correct intervals. If the finishing could be further mechanised, this would increase the saving.

The final factor is the material used. In the future the use of plastics with high rigidity and high dimensional stability could provide cheap, light, non-corroding connection components, with very low maintenance costs.

*Chapter 4*

# Anchorages

## General

An anchorage transfers the forces within the net cables at one or many points into the ground.

The forces acting in the structure are:
(a) constant prestressing load;
(b) long period stresses, e.g. snow loads;
(c) short period fluctuating loads, e.g. wind loads.

Because of the lightweight nature of the structure the self weight load is minimal and therefore the foundations are not, as in heavy structures, in compression, but in tension. The tensile forces caused by prestressing and long period loads are also greatly increased by wind suction loads.

In a conventional tall building resultant wind forces act at a slightly inclined angle to the ground (Fig. 82).

A low rectangular building has a line of force with a slightly larger angle of inclination (Fig. 83). However, on a domed structure the suction force acts almost at 90° to the horizontal. This effect is particularly strong in structures without enclosing walls. Wind gusts can vary from a force equal in weight to 40–200 kg/m$^2$, which is considerably greater than the opposing force created by actual structural weight (Fig. 84). The resulting tensile load of all these forces is to try to pull any foundation from the ground, thus the foundations have to act as anchorage points. The anchorages used fall into two broad categories, anchor plates and friction anchors. Each of those can be designed to oppose either vertical forces or forces acting at an angle to the ground.

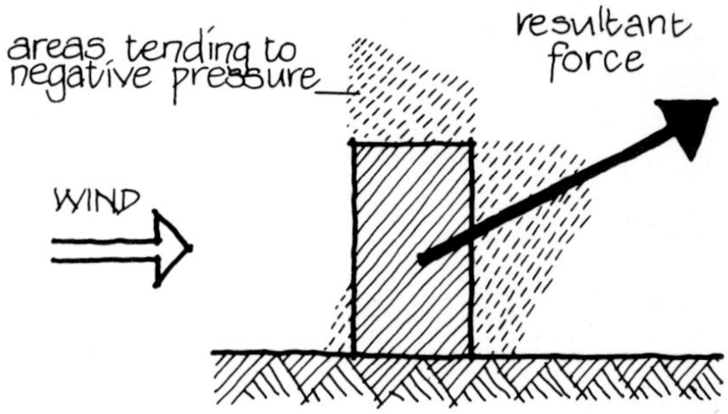


Fig. 82

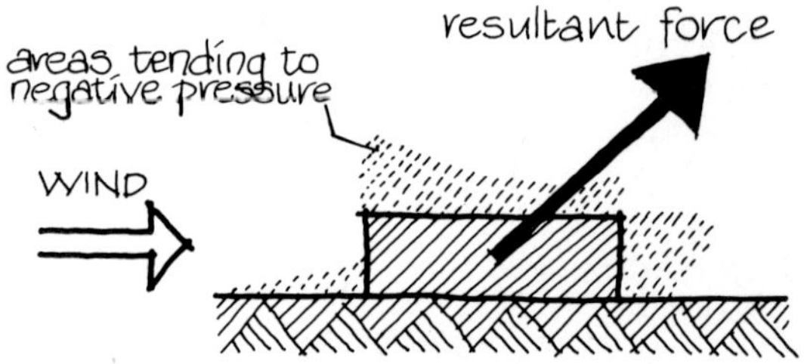


Fig. 83

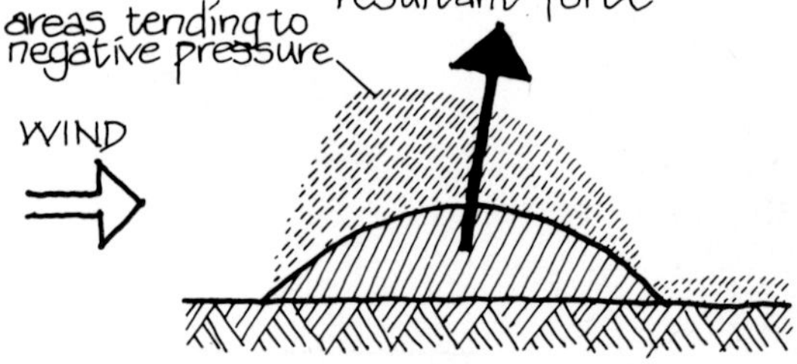


Fig. 84

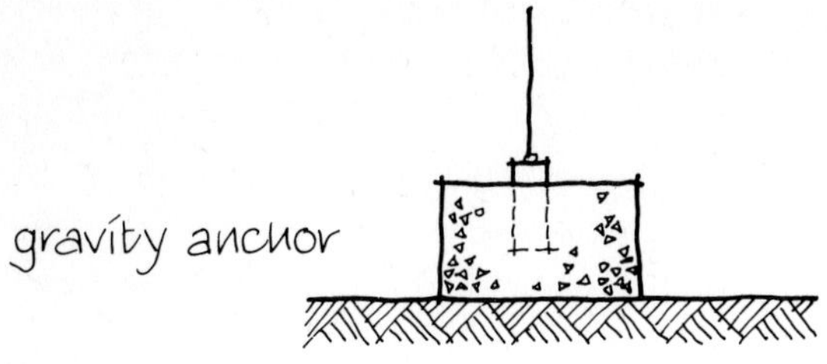


Fig. 85

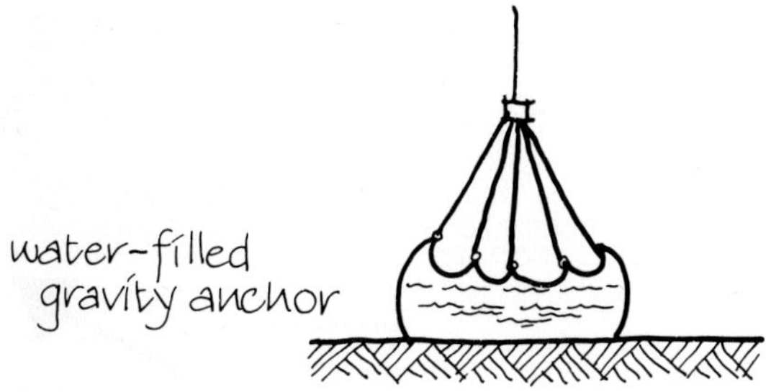


Fig. 86

Fig. 87

Fig. 88

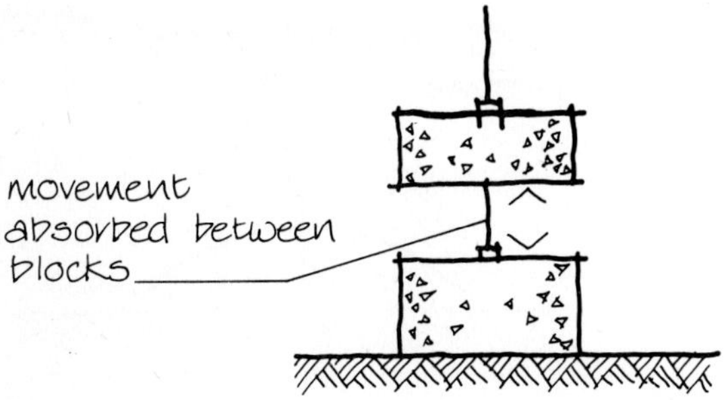


Fig. 89

## Anchorage for vertical tension

### *Simple anchorages*

The most basic type of anchorage is the gravity anchor, which can support tensile loads up to and equal to its own weight when placed on the surface of the ground (Fig. 85).

For mobile lightweight construction gravity anchors of sand, earth or water-filled bags give a resilience which can be used to absorb movement from the structure (Figs. 86, 87).

Mobile lightweight gravity anchors can also be made from concrete, the most basic having a cast in stirrup. The cable is attached to the stirrup using a tension screw (Fig. 88).

A resilient anchor of two concrete blocks can be used to absorb any thermal or moisture movement in the structure (Fig. 89).

By burying the anchor, a greater resistance to load is given, for the weight of the soil above adds to the effective weight of the anchor (Fig. 90).

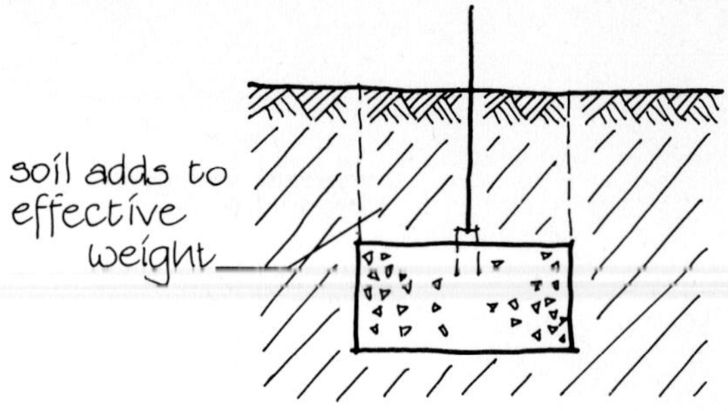


Fig. 90

A buried plate can replace the mass of the anchor (Fig. 91).

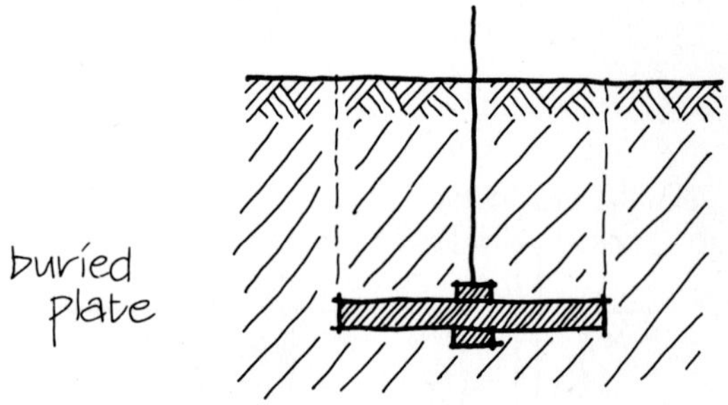


Fig. 91

To keep the resistance to tensile forces the same as in the previous example, the plate must be buried lower into the ground. To increase resistance the plate can be widened or placed deeper still (Fig. 92).

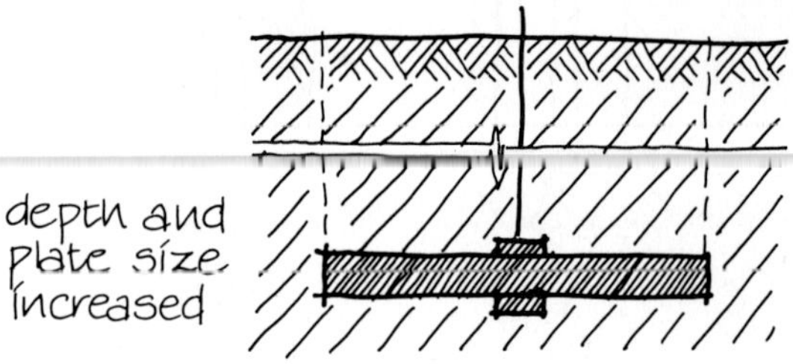


Fig. 92

### *Permanent anchorages*

For permanent large structures friction anchorages are more usual. These rely on the anchor penetrating the ground through a long distance, the rougher the surface and the more solidly entrenched the more resistance the anchor creates (Fig. 93).

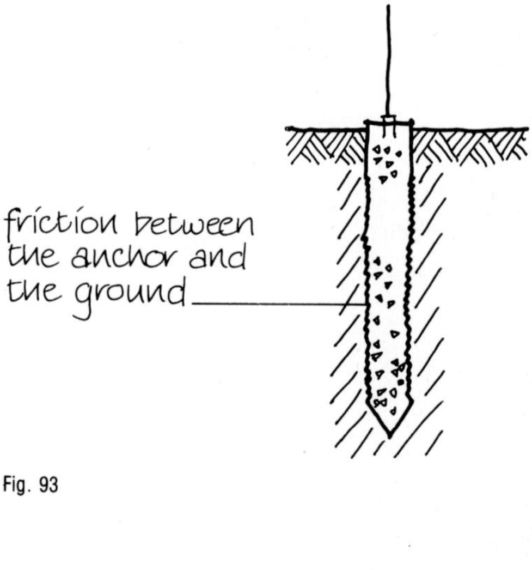


Fig. 93

These are the various types of friction anchor used for differing loading and ground conditions.

*Screw anchors*

These are twisted into the ground like a carpenter's screw (Fig. 94). The anchor should enter the ground as smoothly as possible in order to prevent loosening of the ground around the shaft. The earth should be consolidated around the screwed anchor, thus improving the resistance to tension. If the ground cannot be penetrated without removing the displaced earth then the borehole will have to be carefully tamped back after drilling (Fig. 95). This method is cheap and fast. For small structures anchors screwed in by hand to a depth of 2 m with diameters of 100 to 350 mm can give 1 to 8 tonnes of resistance.

*Convoluted piles*

These piles are driven in, so they have to be strong enough to prevent the shaft from buckling. The shaft itself is profiled in such a way as to cause it to turn as it enters the ground (Fig. 96). Once in the ground, it may have a tendency to unscrew itself. This can be prevented by linking by a yoke

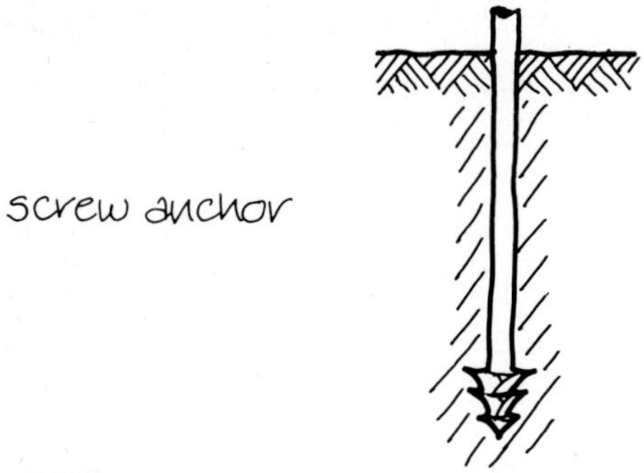


Fig. 94

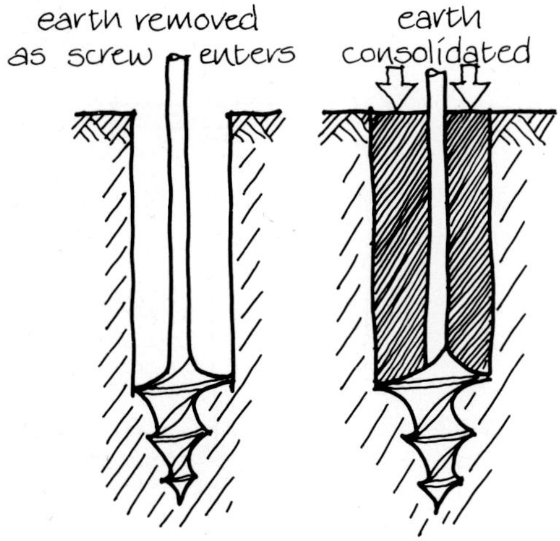


Fig. 95

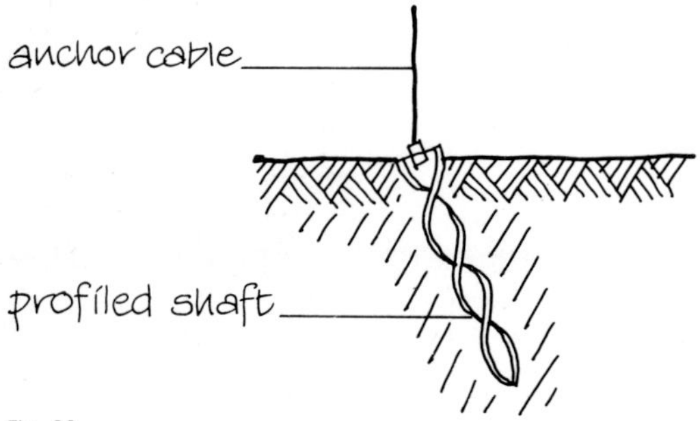


Fig. 96

to another pile close to it (Fig. 97). If the pile is left free to rotate it performs best at a slight angle to the line of tensile force.

The earth around these anchors can be consolidated by driving in dummy piles which pack the earth closer to the shaft (Fig. 98)

*Expanding anchors*
One or more plates are attached by hinges to the shaft of the anchor. This is then driven in to below the required ground depth and then pulled back. As it is pulled back the plates are forced open (Fig. 99). The plates can be situated at the tip of the shaft or at intervals down the shaft itself. The latter case is used if a slim pile is required; having a larger number of smaller plates gives a better resistance (Fig. 100). These 'harpoon' types of anchor once installed are 'lost' because they cannot be recovered without damaging the plates or tip.

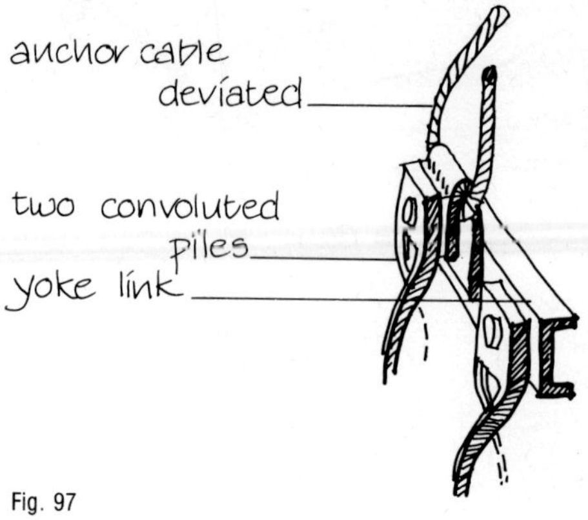

Fig. 97

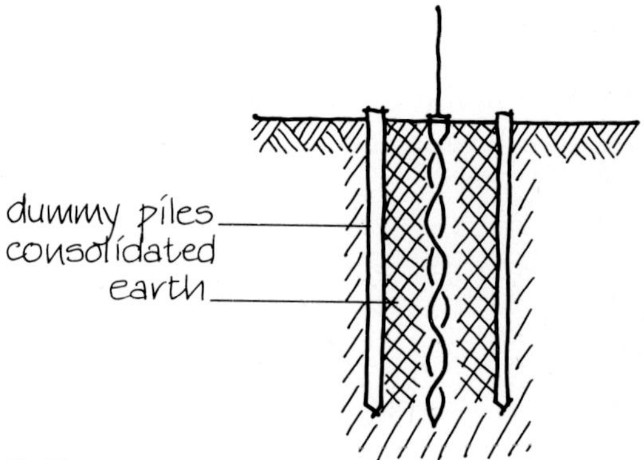

Fig. 98

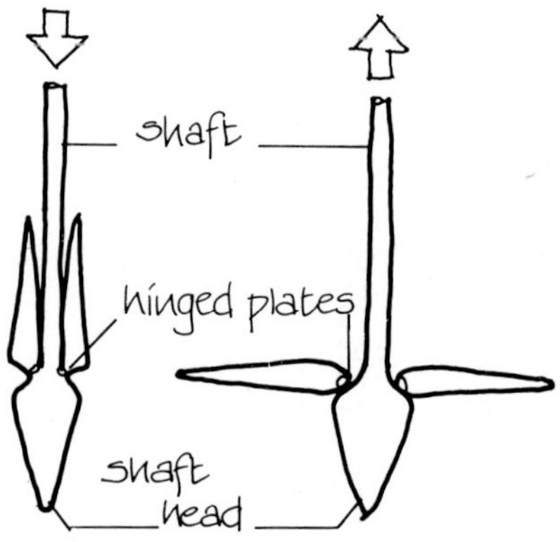

Fig. 99

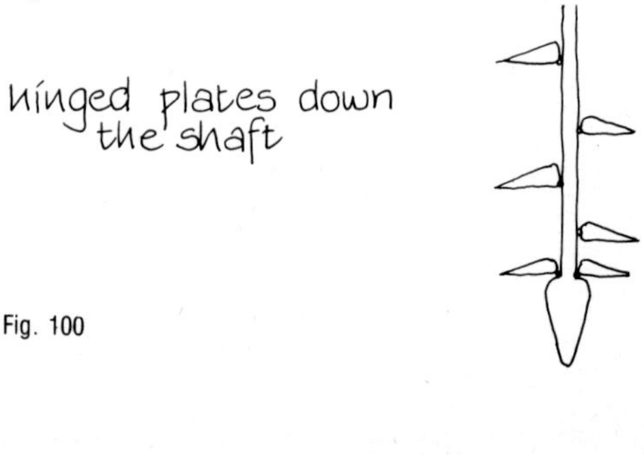

Fig. 100

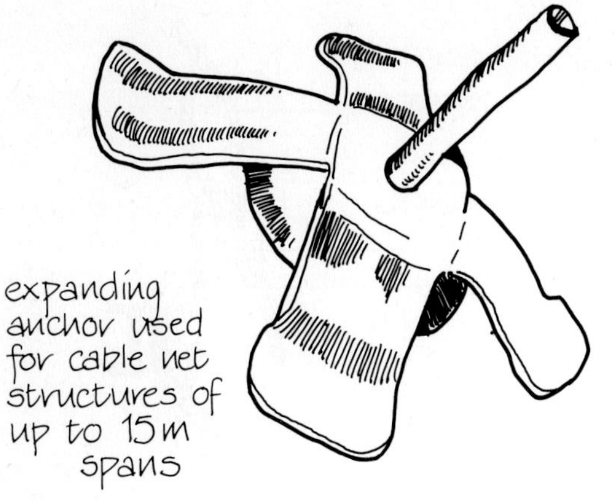


Fig. 101

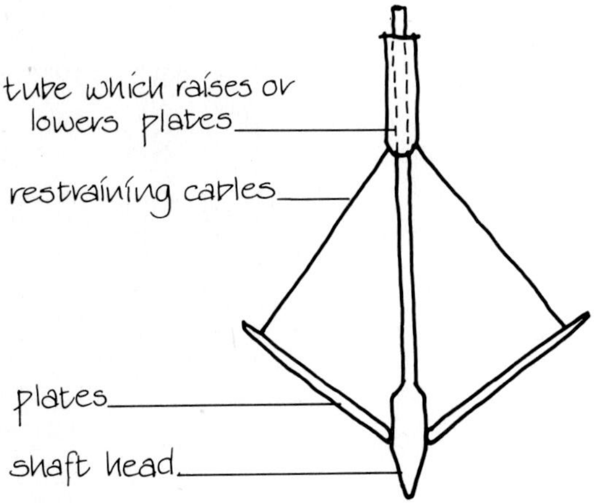


Fig. 102

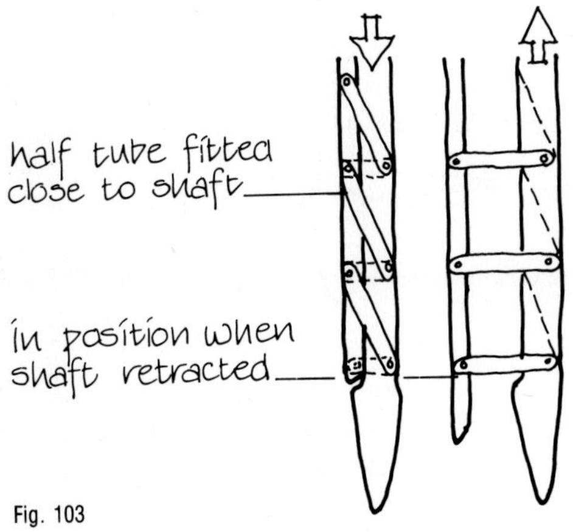


Fig. 103

Most anchors are inserted in a pre-drilled bore hole. An anchor with four leaves which fits a 250 mm diameter bore hole has a surface area of 1,200 $cm^2$ when expanded. In sand this gives a resistance of 6 tonnes, in sandy clay 8 tonnes and in hard soil over 15 tonnes. These anchors can be driven by one person and are commercially available and consequently are reasonably priced (Fig. 101).

*Anchor needles*
These are another form of anchor with an expanding tip. The tip can either expand using retractable plates (Fig. 102), or the tube can be split longitudinally near the tip. The separate section is attached in such a way that, on attempted withdrawal of the anchor, it swings outwards and jams into the hole (Fig. 103).

Anchor needles which use a 'lost' tip have a tip which is expanded to give the necessary resistance to load. Explosives or compressed air and water can be used to expand the tip (Fig. 104).

*Bored footing piles*

These act in a similar way to expanding anchors in that the tip is enlarged and acts to increase the ability to take tension loads.

A bore hole is drilled into the ground and a space for the anchor foot is cut using a steel loop. The cavity is then lined with a steel tube and a tension cable is introduced. Concrete is then poured in and tamped to fill the footing and tube (Fig. 105).

*Folding anchors*

A folding anchor consists of a tension member, either a rod or cable, which is introduced into the ground; this has several folding plates, flaps or ribs attached to its base. When in position these are folded out to stabilise the anchor. In one system a tube is bored into the earth, a folded anchor on a cable is placed in it, the tube is withdrawn and the anchor is then unfolded by pulling the cable upwards. The bore hole is filled with concrete to finish (Fig. 106).

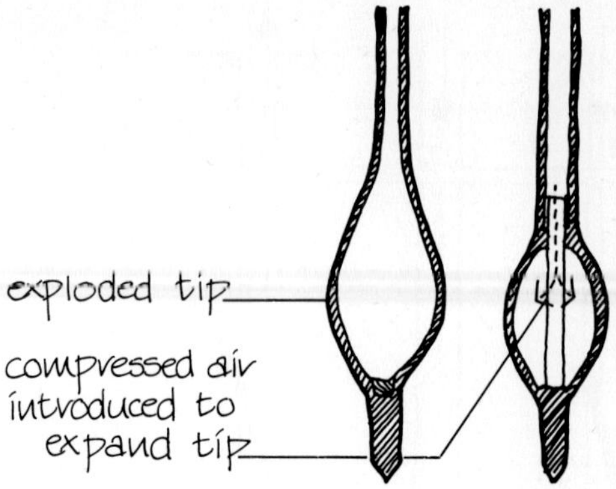


Fig. 104

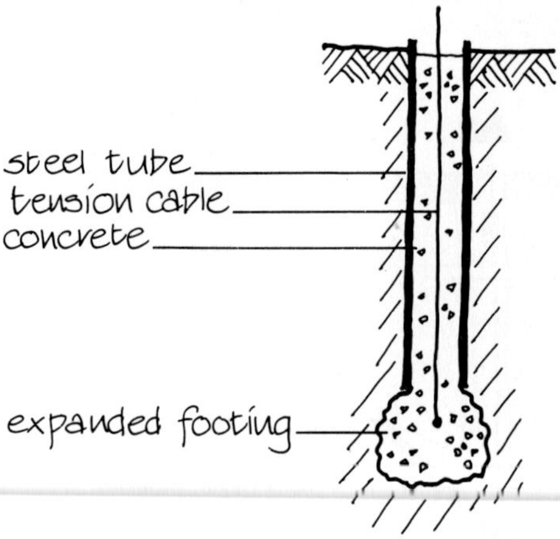


Fig. 105

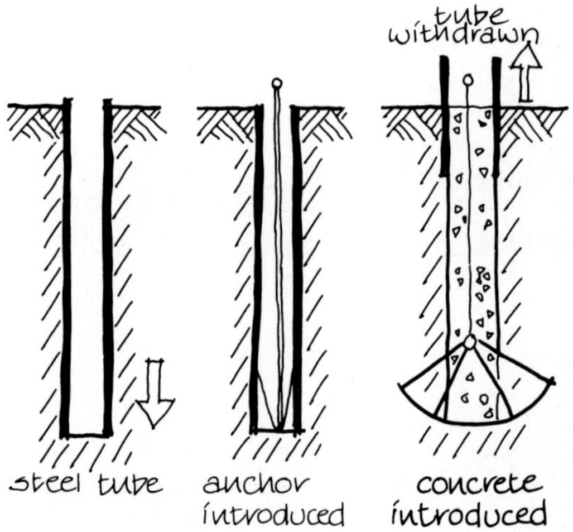


Fig. 106

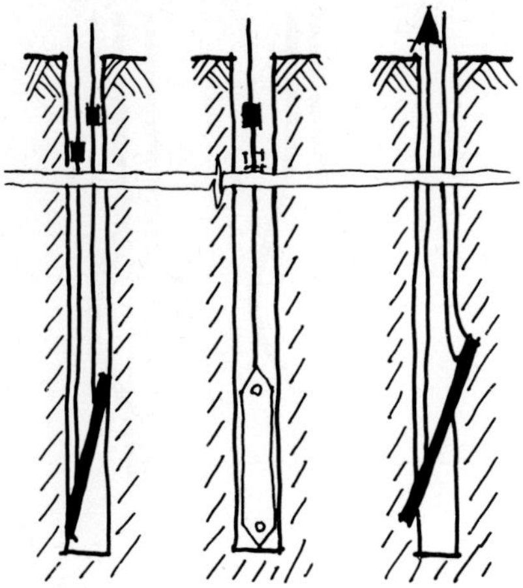

Fig. 107

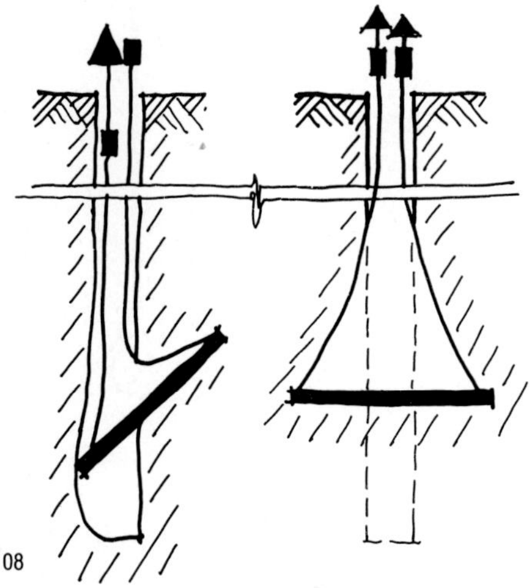

Fig. 108

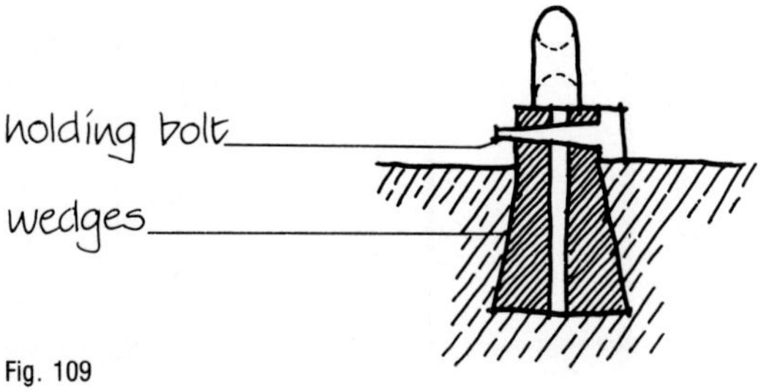


Fig. 109

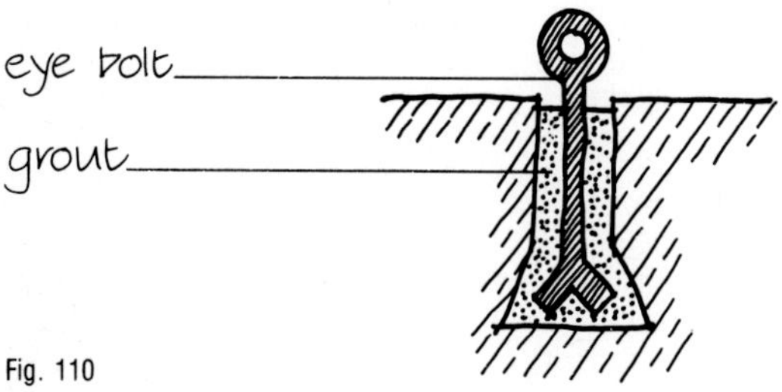


Fig. 110

For anchorages experiencing an exceptionally large tensile load, a safety element is achieved by the use of two cables as tension elements. A bore hole is made and an anchor plate with a cable attached to either end is placed in it. The lower cable is tensioned and this pushes the upper end of the plate into the soil (Fig. 107). Markings on the cables show when the plate is horizontal and indicate the depth of the plate in the ground (Fig. 108).

The cables are tightened by wrapping them round a bar attached to an oil pressure cylinder and a spindle. By using this instrument the anchor is brought into position and the resistance of the soil is registered.

This type of anchor has the advantage that it is recoverable by releasing one cable and pulling on the other; the plate is thus retrieved.

*Anchorages for solid ground*

Anchorages in solid ground which experience tension in a vertical direction only can use special grouted-in members; these are suitable for use in rock, which will withstand relatively high tensile loads.

A simple recoverable anchor, known as a 'wolf', consists of two wedges placed in a swallow-tailed shaped hole in the rock and wedged into position to hold the anchorage in a central position. The anchor is recovered by loosening the holding bolt on the surface (Fig. 109).

Simple grouted anchors may be formed by eyebolts and threaded bolts grouted into holes which are wider at the base than at the top (Figs. 110 and 111). The best grout to use is one which expands slightly on setting. These simple anchors are best used on small projects on rocky ground.

*Grouted anchors*

For high tensile loads a cable tied at its end is introduced into a deeply drilled hole. The tightening of the cable causes it to widen slightly under its own weight. It is then grouted in using cement pumped at high pressure (Fig. 112). If the cable is passing through the concrete to a great depth it is best to isolate the upper part of the cable with a bitumen coating. This prevents the cable under tension acting with the concrete and so avoids the concrete taking any tensile loads.

This system can take very high loads in solid rock and also in medium strength granular soil, if the grouting is firmly embedded.

*Compression piles*

Piles which have been developed for use in compression can be used in granular soils to take tension. A bore hole is made and a tube inserted; the concrete is tamped into the hole as the tube is withdrawn so ensuring that the concrete can form a good bond with the ground (Fig. 113).

The concrete will tend to spread further into soft layers than into hard layers of earth. These piles designed for compression have steel reinforcing to cope with any compressive forces in the concrete and a tension member at the centre. The tension cable may be anchored only at the foot and the remainder of the cable isolated from the concrete by means of a bitumen or other coating. Alternatively, it may be cast into the concrete but only if the concrete is precompressed to a level that is greater than the maximum tension predicted. This prevents the concrete from cracking if placed in tension.

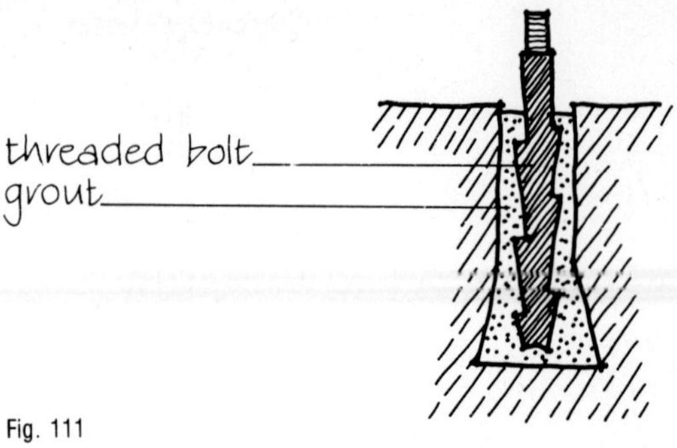


Fig. 111

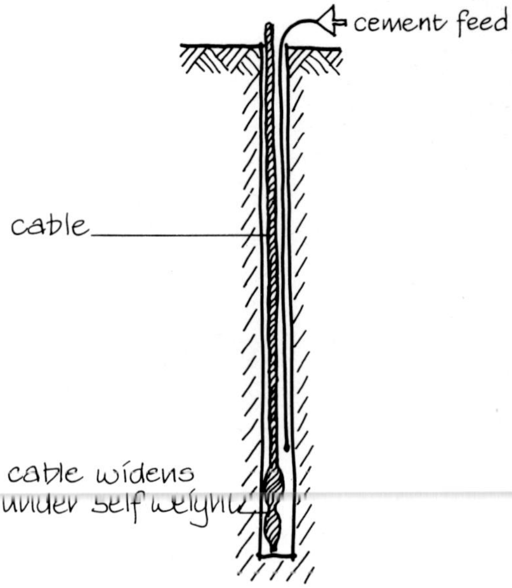


Fig. 112

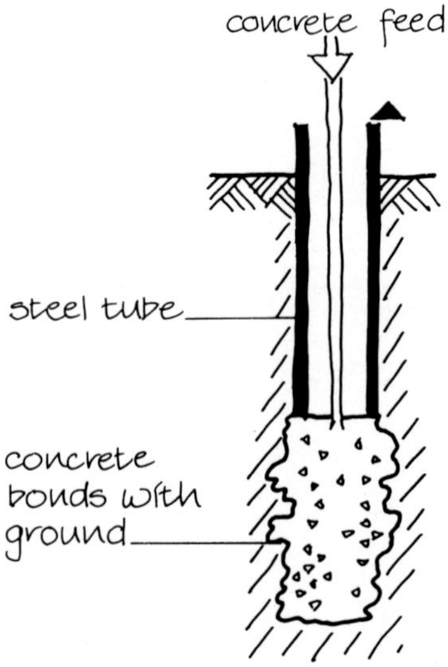


Fig. 113

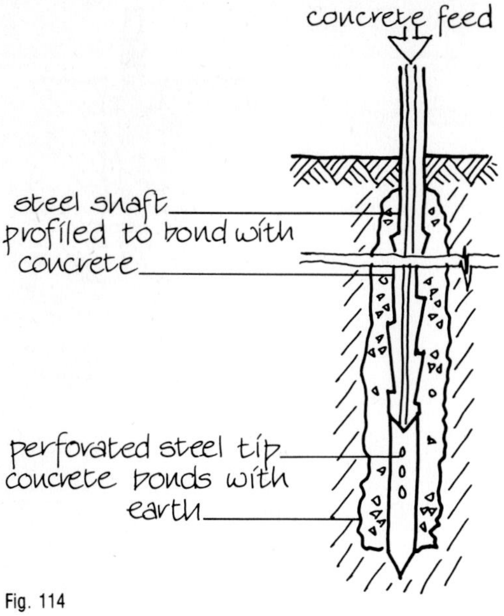


Fig. 114

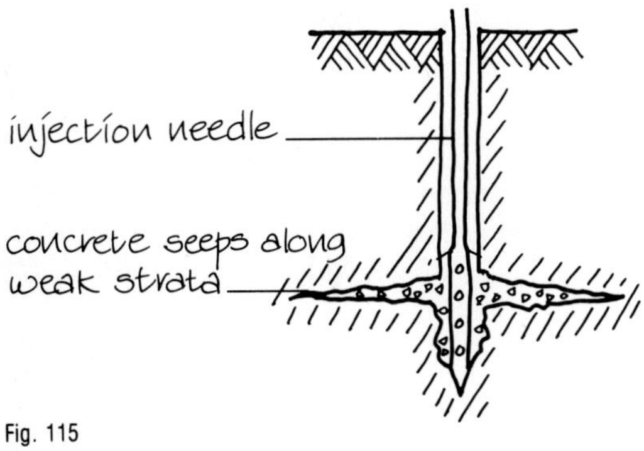


Fig. 115

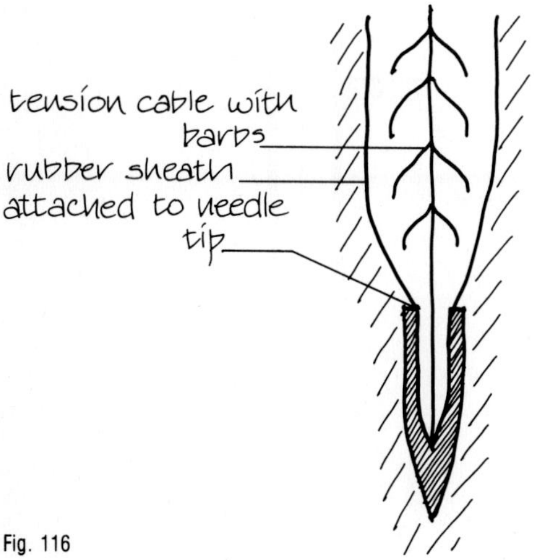


Fig. 116

*Injection anchor needles*

Injection anchor needles were developed for use in all soils for very high loadings. A thin steel pipe with a perforated tip is driven into the ground, then the concrete is forced down the needle and through the perforations. The concrete sets in the soil and acts to consolidate it around the needle (Fig. 114). To drive injection needles which can be very thin and have little resistance to buckling the needle must be driven from just above the ground using clamping jaws or, in sandy earth, water at high pressure.

In uneven soil a special needle must be used to prevent the concrete seeping only into the softer layers of ground; this would leave weak bands where the soil was not consolidated (Fig. 115). To overcome this problem the needle is enclosed in a soft rubber or plastic case, which can expand into the ground when the cement is forced in at pressure (Fig. 116). The rubber sheath prevents the grout from escaping into the soft layers and so the anchor is stabilised.

## Inclined anchors

If the angle of inclination of the line of force is equal to or less than 30° to the horizontal an anchor designed for vertical pull cannot be used (Fig. 117).

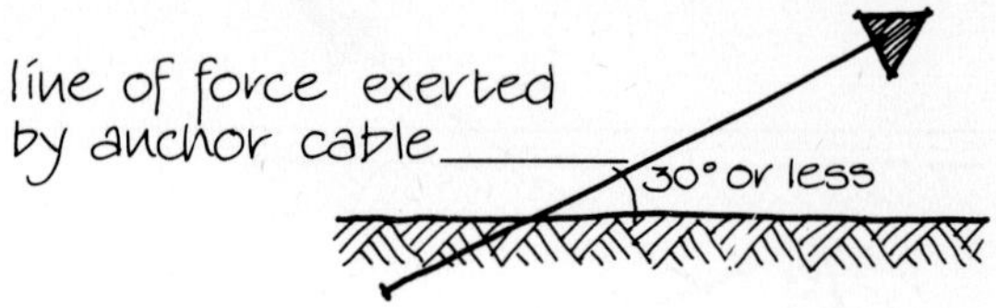


Fig. 117

### *Gravity anchors*

Gravity anchors rely on friction with the ground to withstand the forces acting on them (Fig. 118). To increase the loads which can be taken by a gravity anchor the friction between the ground and the anchor must be increased. The base of the anchor can be serrated (Fig. 119) or the block can be held by a pile (Fig. 120) or the block can be buried. The frictional forces will be increased due to the extra weight of soil above the block and the extra surface area in contact with the ground (Fig. 121). The resistance will be greatly improved if the concrete forming the block is tamped directly into the trench as this will make the bond between the earth and the concrete firmer.

In a buried gravity anchor it must be remembered that the front of the block will be in high compression due to the resistance of the ground (Fig. 122).

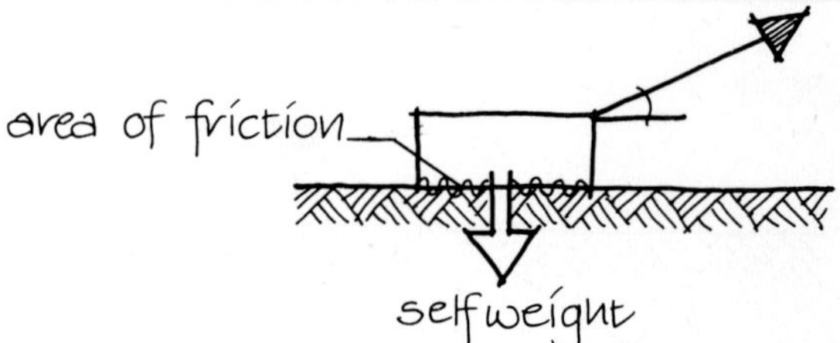


Fig. 118

Fig. 119

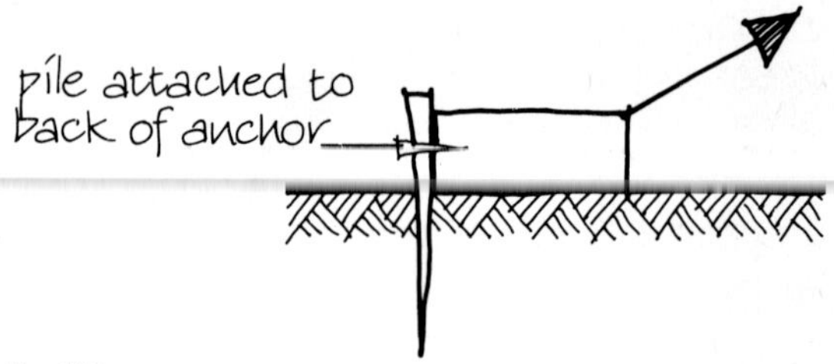


Fig. 120

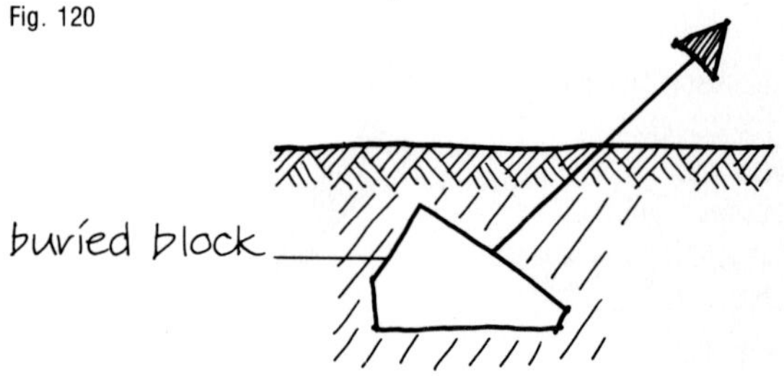


Fig. 121

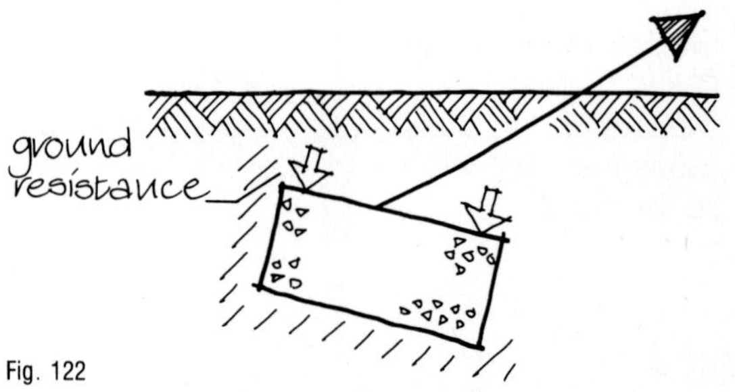


Fig. 122

THE LIBRARY
GUILDFORD COLLEGE
of Further and Higher Educati

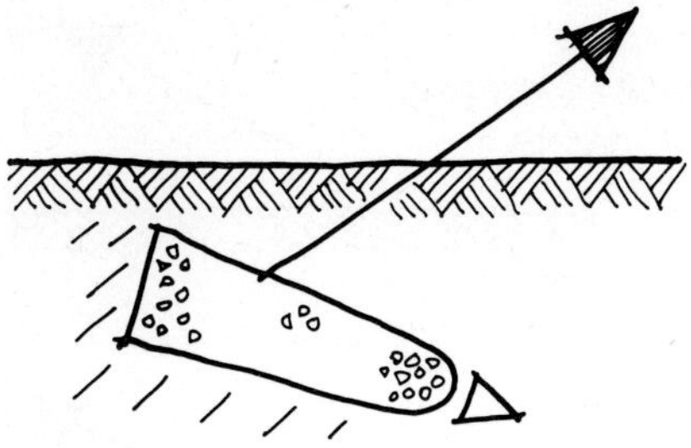

Fig. 123

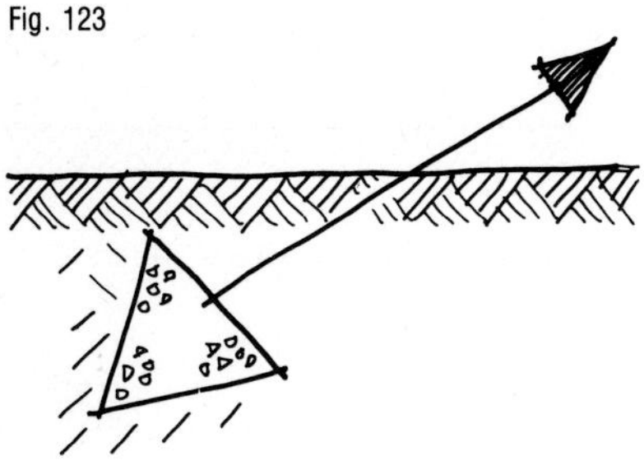

Fig. 124

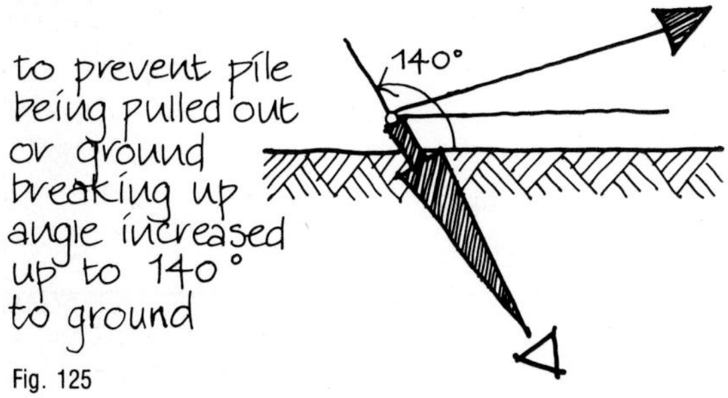


Fig. 125

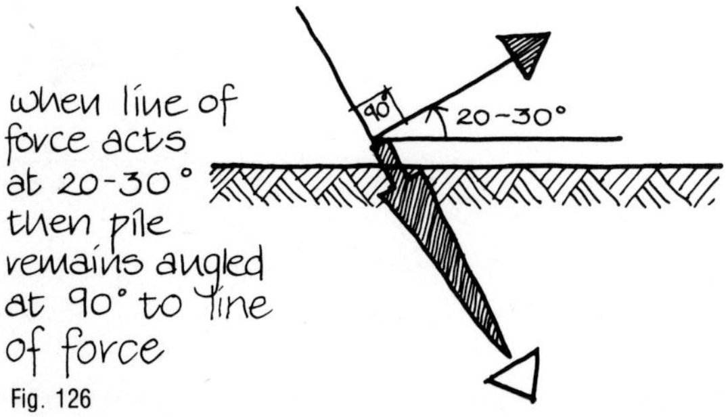


Fig. 126

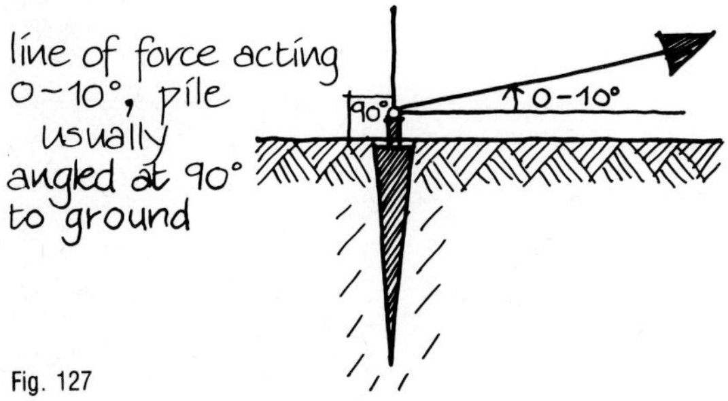


Fig. 127

A wedge-shaped block will act as a ploughshare and dig itself into the earth as the cable is tightened (Fig. 123).

To give the maximum resistance using the minimum of material a triangular anchor can be used (Fig. 124).

### *Friction anchors*

*For good soil*
Piles driven into the ground can be used to anchor inclined loads. The pile is usually driven in at an angle; this is to prevent it being pulled out or breaking up the ground in front of it (Fig. 125).

When the pile is at an angle it will tend to act in a ploughshare manner and penetrate deeper into the earth under increasing load.

The pile is usually angled at 90° to the line of force except when the angle of inclination of force is between 0–10° to the horizontal (Figs. 126 and 127).

The type of anchors mentioned in this section can be used on the incline and can resist larger loads than the inclined gravity anchors.

*Friction anchors in poor soil*

If the cohesive quality of the soil is poor it is more satisfactory to arrange the tension element to be carried by a number of smaller angled piles. This gives the soil less chance of breaking up in front of the pile. The piles can either be arranged as a pile wall (Fig. 128) where they are joined together using a steel plate, or the piles can be fanned out (Fig. 129). The piles, if overloaded, will give slightly until the tensions in all the cables are equalised. To increase the resistance of the ground to breaking up, dummy piles can be driven between the fanned piles in order to consolidate the ground around the piles (Fig. 130).

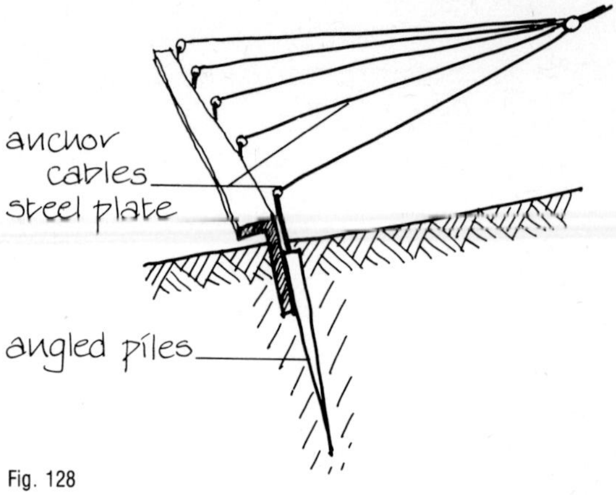


Fig. 128

## Non-structural considerations

### *Protection*

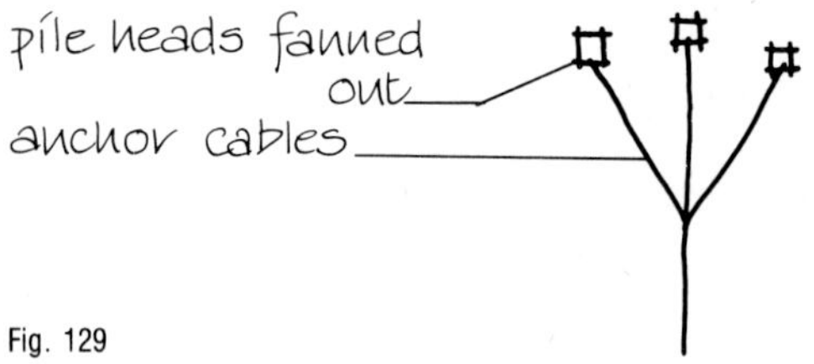


Fig. 129

The ground conditions dictate to a large extent the degree of protection against corrosion necessary for the anchorage. Such factors as the acidity of the ground, the sulphur content and the water table level must be considered in the design of any concrete work within the anchorage. To avoid frost damage the concrete must be buried at least 750 mm below ground level; this applies particularly to gravity anchors.

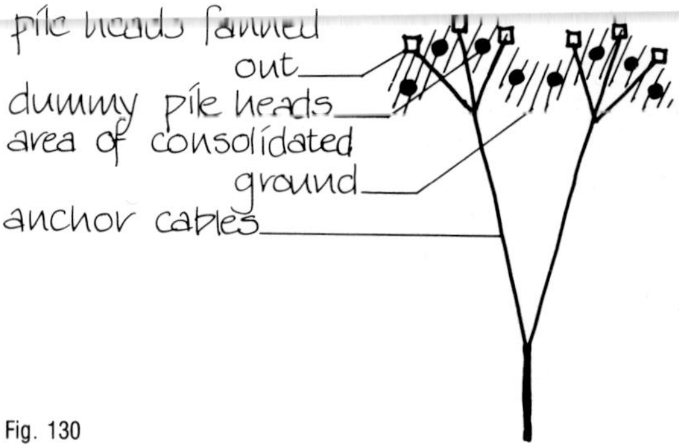


Fig. 130

The other element found within a tension friction anchorage is the tensile cable or rod, this usually being made from a metallic filament, such as steel. As steel corrodes rapidly in the presence of water, steps must be taken to ensure that any steel work within the ground is adequately shielded from ground water. This is usually achieved by encasing the tension cable with bitumen and concrete to protect it.

### *Cost implications*

The factors which affect the cost of the anchorages in comparison to the overall costs of a project are the load conditions, the required length of life of the structure (whether considered temporary or permanent) and the ground conditions.

The first factor affects the total cost of the anchorages because if the loads are increased then the anchorage has to withstand higher tensile forces. This necessitates an increase in the amount of material used or the use of more complex techniques in order to give the anchor better holding power in the ground.

Short-term structures have anchorages which do not remain in the ground for long periods, therefore the problem of preventing corrosion of the anchor does not occur. Smaller and more simple unprotected anchorages can be used, especially as temporary or movable structures also tend to be subject to smaller loads.

Permanent tensile structures have to transmit much higher forces and the anchor remain in the ground for long periods of time, therefore they have to be much stronger and well protected. This of course increases the cost of the anchorage.

When the bearing capacity of the ground is high then it becomes easier to transfer the loading of the structure to the earth. This in turn means less material and less complicated anchorage devices which provide an economy on the total cost.

However, the percentage cost of the whole of the anchorage remains approximately the same despite these varying factors. This is because as the costs of the anchorage increase with, for instance, an increase in loading, so do all the other costs in the structure.

Also in the case of long or short-life structures, the cost of long-life anchorages is much higher than those of short-life ones, but the costs of all other components in long-life structures are proportionally increased.

The independent factor, which can alter the balance of anchorage costs to overall costs, is the nature of the ground; high-bearing capacity ground reducing the percentage costs, poor ground increasing them. Good design can also reduce the percentage costs. As the usual costs of the anchorages is between 20–25 per cent of the total cost, any savings made here can be valuable.

*Chapter 5*

# Supports

## General

A support is the structural element which fixes the position of the cable net in space in relation to the ground. The support acts together with the anchorages to counterbalance the tension created in the cables. It can be positioned internally or externally in relation to the net. It can be solid or skeletal and classified into the following categories:

- One dimensional, e.g. columns and arches (Fig. 131).
- Two dimensional, e.g. slabs and shells (Fig. 132).
- Three dimensional, e.g. space frames and branched columns (Fig. 133).

The most important of the three support groups, because it is the most commonly used, is the one-dimensional support. This group includes masts (columns) and arches. The support is an element which functions chiefly in compression; its inherent characteristics are therefore that the member uses a large amount of material and takes up a large volume in order to transmit the loads applied to it, unlike the element in tension, such as a cable net, which is highly efficient in its use of material (Fig. 134).

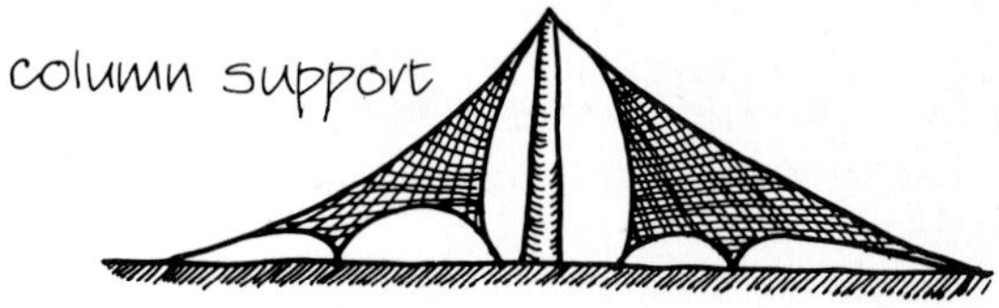


Fig. 131

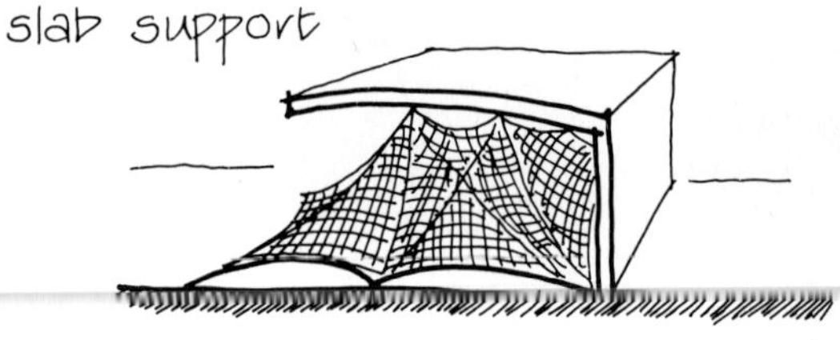


Fig. 132

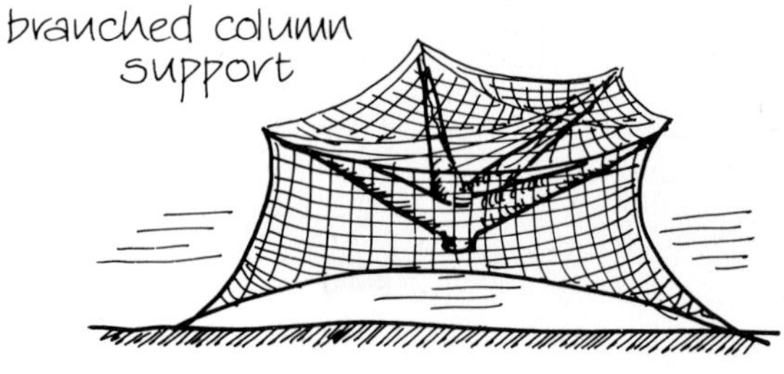


Fig. 133

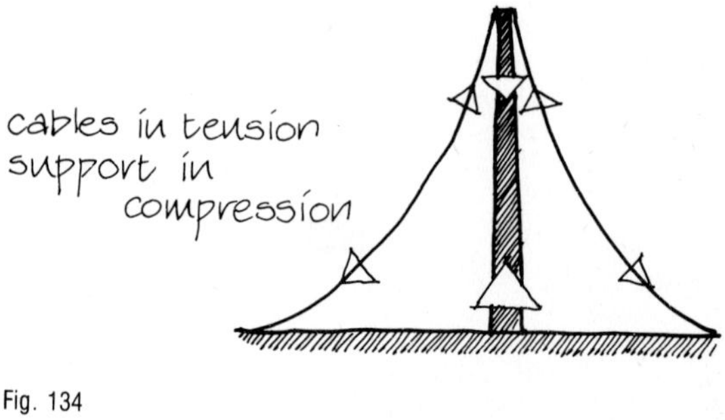


Fig. 134

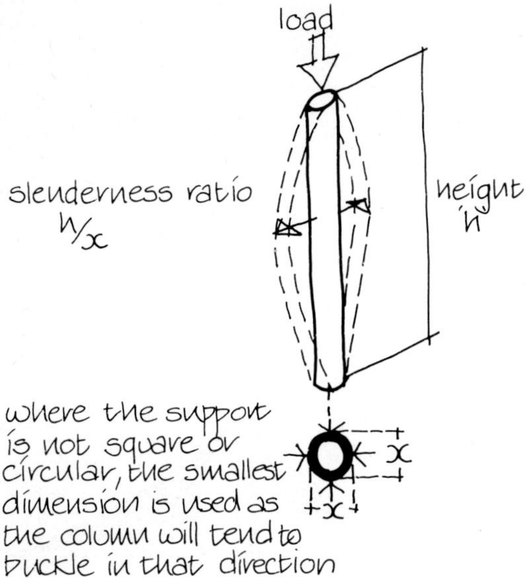


Fig. 135

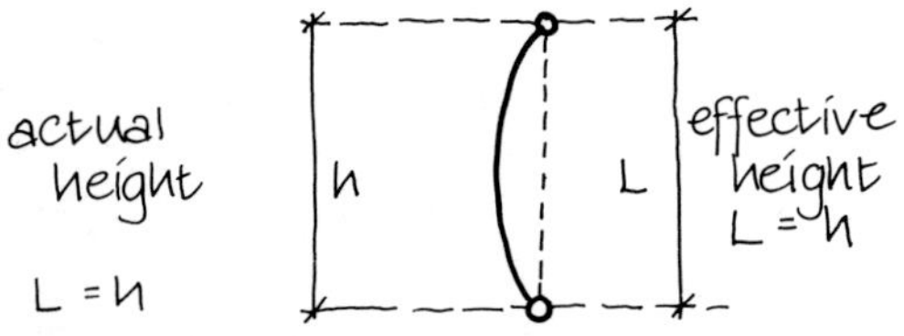


Fig. 136

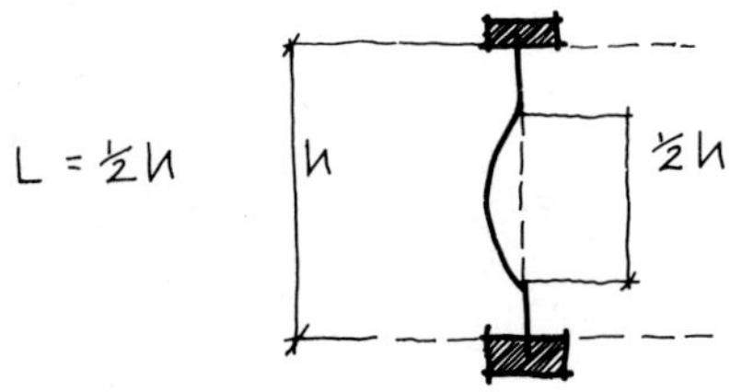


Fig. 137

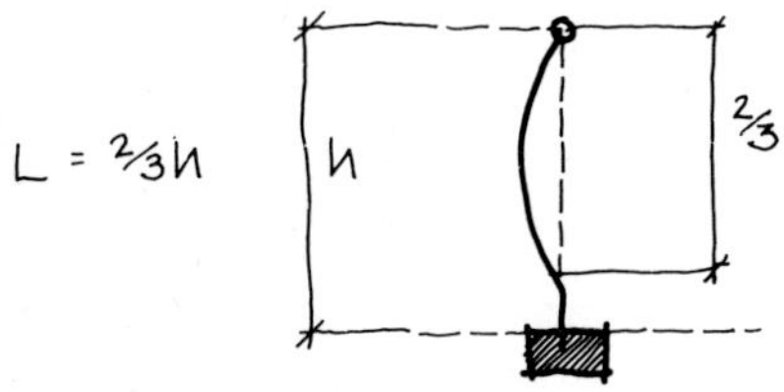


Fig. 138

It is necessary for the support member to be able to withstand the forces which are applied to it by the cable net. The tendency of a support which is tall, that is defined as having a height to thickness ratio (slenderness ratio) of 15 or more will be to buckle in some lateral direction (Fig. 135).

The end fixing conditions will affect the way in which a support will tend to buckle. If it is fixed in position but free to rotate or pivot at both top and bottom it is 'hinged' and free to buckle over all its length (Fig. 136).

If the support is held rigidly in position at both ends (fixed ends) the buckling length is reduced to half its actual height (Fig. 137).

Alternatively, the column can have one fixed end and one hinged. In this case the effective length becomes two-thirds the actual height (Fig. 138).

Thus the maximum load which a support can carry is dependent on its effective height and cross-sectional area.

## One-dimensional tubular masts

### *Hollow steel masts*

One of the simplest ways of providing a support is to use a tapered hollow steel tube for short masts (Fig. 139). These are probably the most economical mast form as the diameter does not have to be excessive in order to prevent buckling. To increase the height of the masts (Fig. 140) without having to increase their external diameter it is necessary to increase the thickness of the shell by decreasing the internal diameter or by bracing the centre of the mast with fins (Fig. 141).

Connections can be easily welded or bolted onto the mast heads for provision of terminals for the cable nets. The base of the mast can be connected either onto a pivoting joint, which in turn is embedded in a concrete foundation or through a fixed baseplate, again to a concrete foundation. It must be remembered that the foundation for the mast will be mainly subjected to compressive forces but under extreme wind loads the suction imposed on the structure could be large enough to place the support in tension.

The mast head must be designed to provide the connections necessary for the cable network of the structure. Some examples are shown in Ch. 3, Connections (deviation and branching knots).

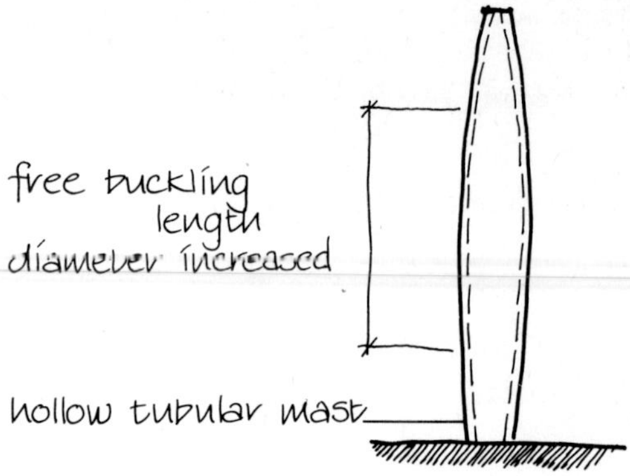


Fig. 139

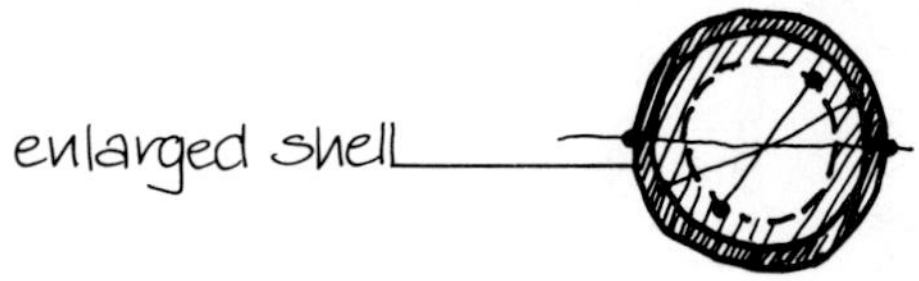


Fig. 140

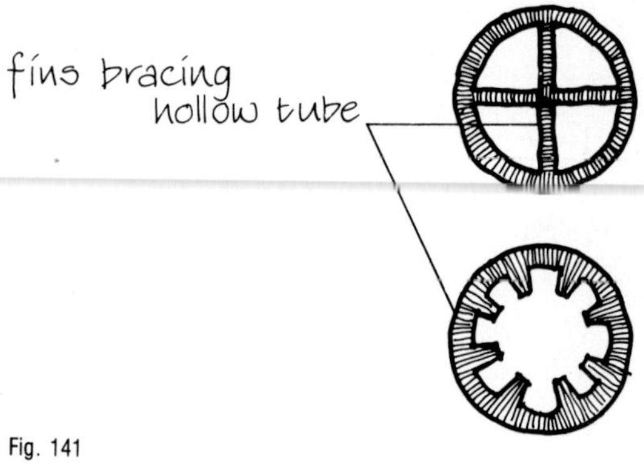


Fig. 141

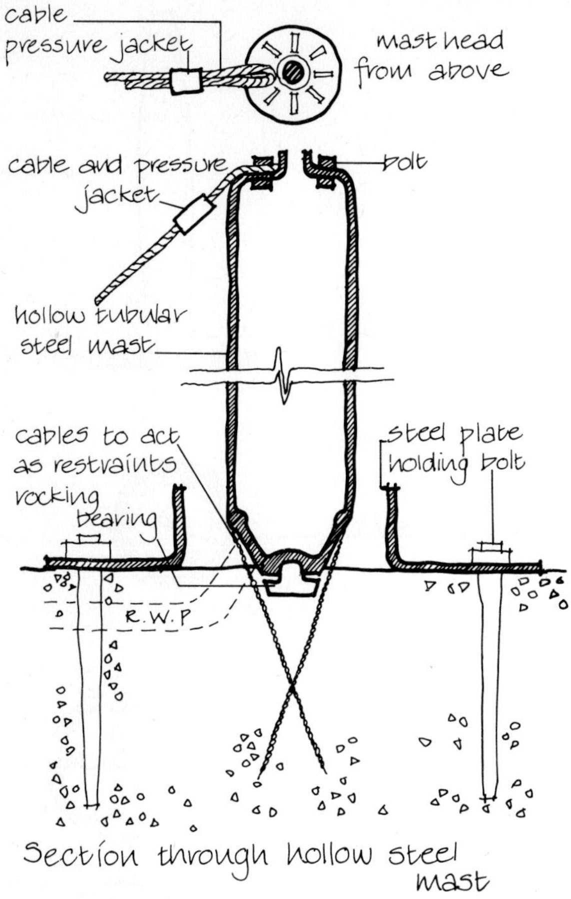


Fig. 142

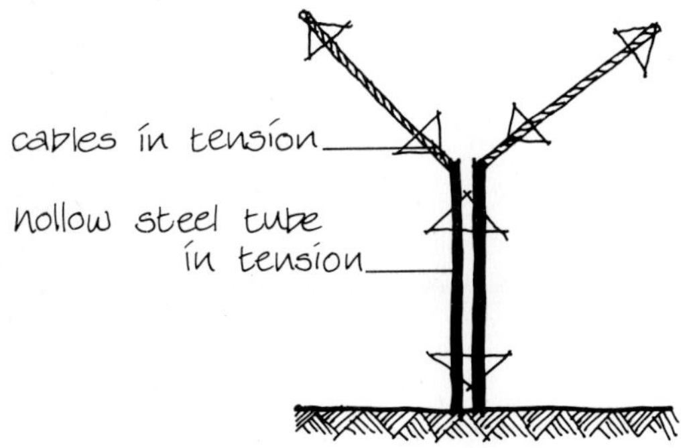


Fig. 143

Figure 142 shows details of a typical hollow tubular mast suitable for use on small scale, perhaps, temporary, projects. The net cables are taken round a bolt or disc which is fixed to a cast steel head plate which has been welded to the top of the mast. Pressure jackets are applied to the cables as termination devices.

The base of the mast is closed and sits on a rocking bearing; this is set onto a steel plate which in turn is embedded in a large concrete foundation.

### *Low point restraints*

When the support is maintaining the cable network in a depressed position it becomes necessary for it to withstand only tension which is caused by the upward pull of the restrained cables, rather than a compressive force (Fig. 143).

If the low point is held in position by a hollow tube support or by tension cables it can also be used as a rainwater drainage channel for the main structure surface (Fig. 144).

Depending on the net covering, fabric strengthening patches may be bonded onto the membrane or a sheet metal funnel can form the transition from the roof cladding to the rainwater discharge pipe. A typical low point restraint uses cladding fixed to circular ring tube which is then connected to a conical assembly of rods. These are welded to the water discharge funnel. An internal rubber water discharge tube runs down the hollow restraint. An overflow funnel forms a safeguard if any leaves or ice start to block the discharge pipe. If the low point has a large opening the overflow pipe is not necessary (Fig. 145).

## Other one-dimensional supports

Arches and point suspensions with cables also constitute one-dimensional supports (Fig. 146). Net cables may be connected to the arch, which acts as the compression member (Fig. 147). The arch can be constructed from reinforced concrete,

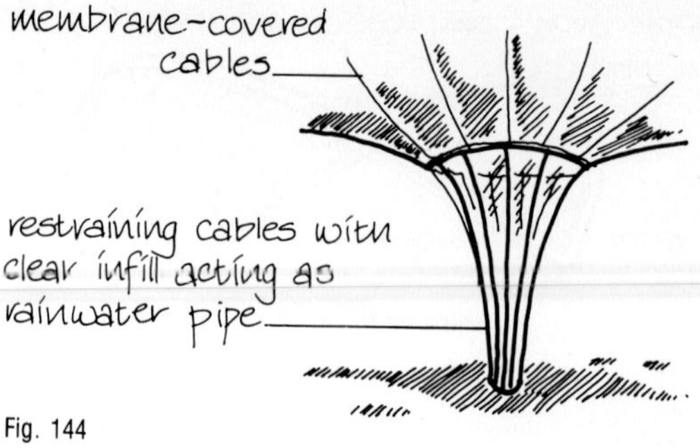


Fig. 144

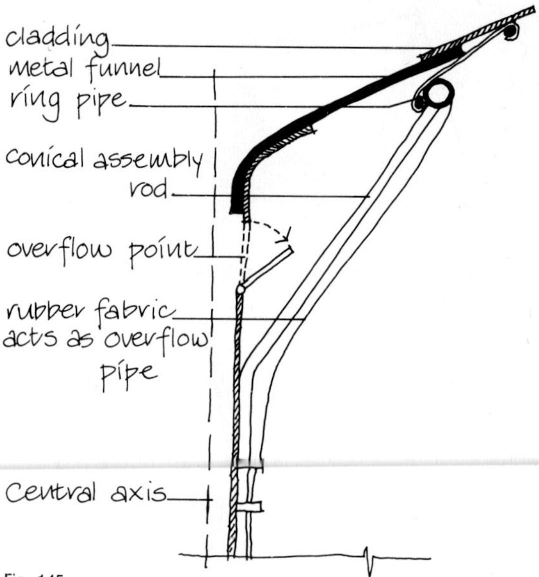


Fig. 145

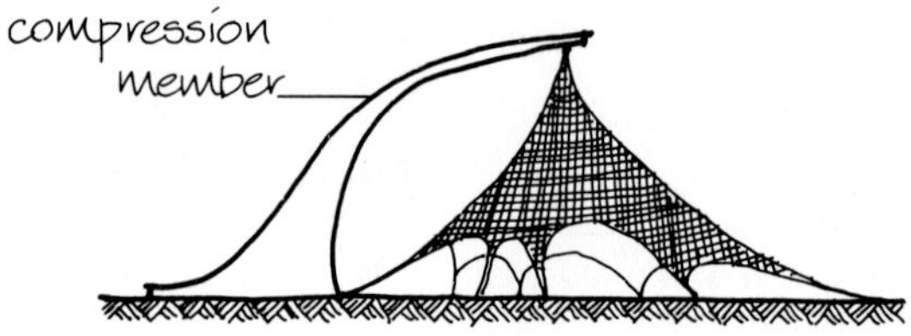


Fig. 146

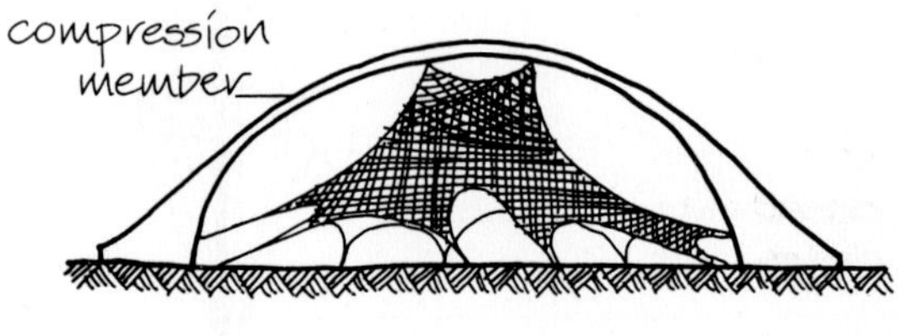


Fig. 147

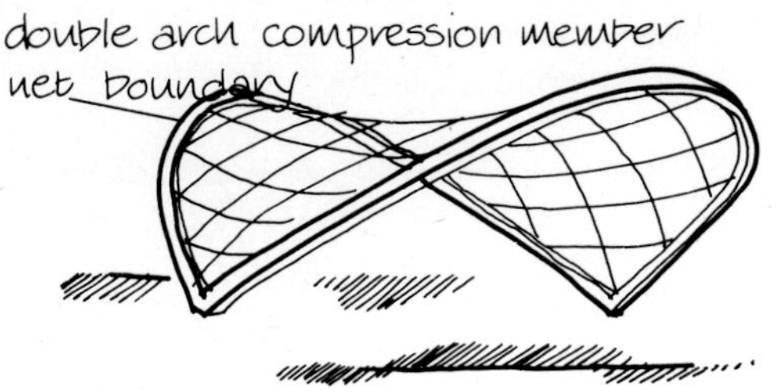

Fig. 148

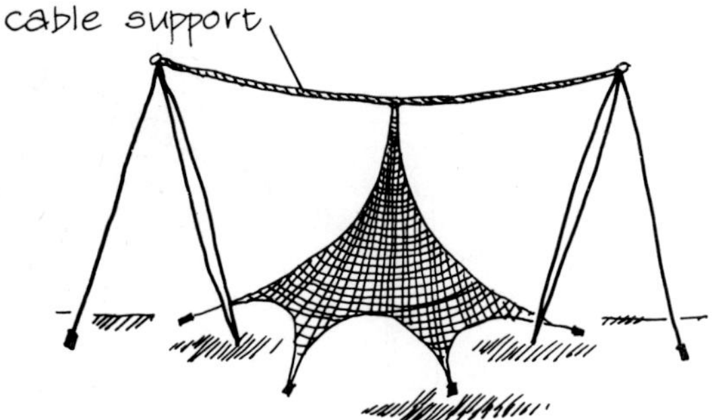

Fig. 149

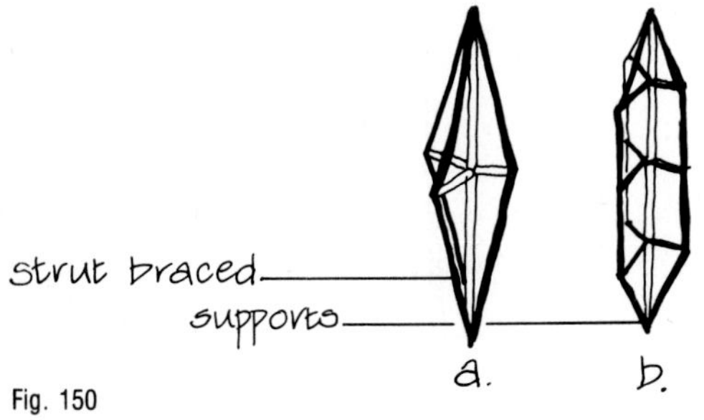

Fig. 150

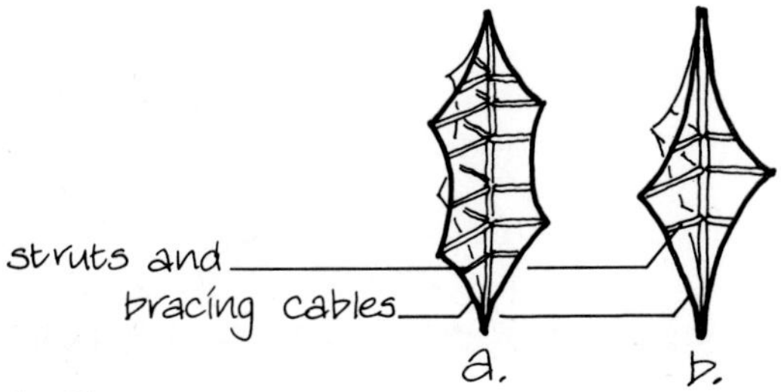

Fig. 151

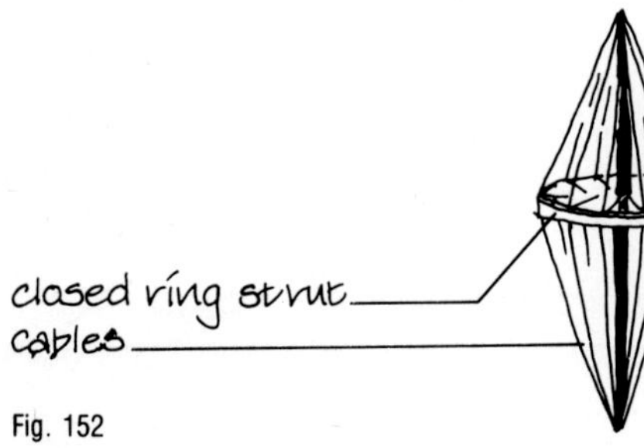

Fig. 152

tubular steel or laminated timber, which all take compressive stresses well. Arches can also be used to form the edge condition of the net (Fig. 148).

The main cable net can be suspended from a cable which in itself is suspended from a support. This combines the simplicity of the hollow mast with the provision of a totally free floor area (Fig. 149).

### *Trussed compression members*

Since the compressive strength of a tall mast is dependent upon its slenderness ratio, either a large mass of material or the introduction of intermediate bracing is required in order to prevent buckling. A trussed compression member will use less material than a thick walled type of large diameter mast, but may require more complex manufacturing processes.

The bracing can be compressive, using short struts (Fig. 150) or a combination of compression struts and tension cables (Fig. 151). When the compressive forces are taken in a bracing system which is closed, such as a ring or triangle, then the strut can be braced entirely with cables (Fig. 152).

The compression pole may be incorporated into cable networks (Fig. 153a), or between membranes (Fig. 153b), but whatever system is used the pole must always be braced in at least three planes.

The compression member may be continuous (Fig. 154), or hinged in intermediate positions (Fig. 155a,b); this increases the flexibility of the member.

The trussed masts most often used in the design of supports for cable nets are those using bracing struts; these give a high performance under compressive load and use the minimum of material.

More than one compression member is often used; these are joined by members acting in tension (Fig. 156). The compression members must all act at the same distance from the central axis of load of the support. They combine at the mast head to give a bearing point for the net cables.

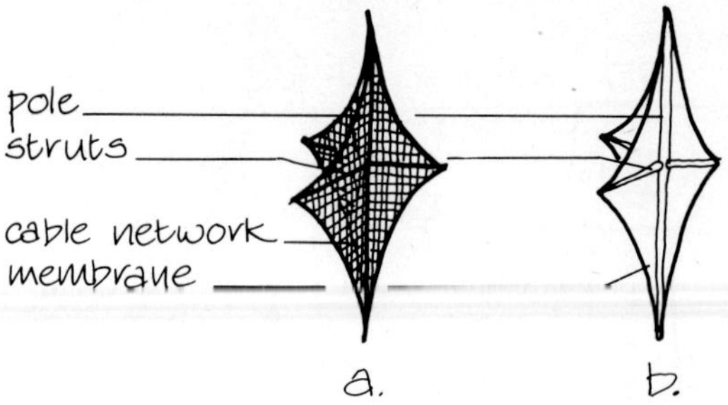

Fig. 153

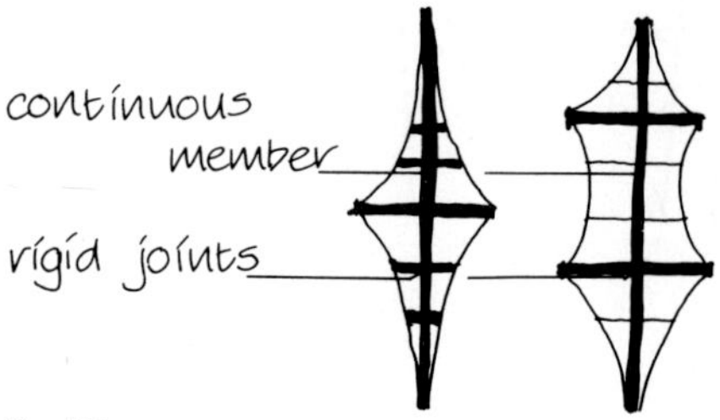

Fig. 154

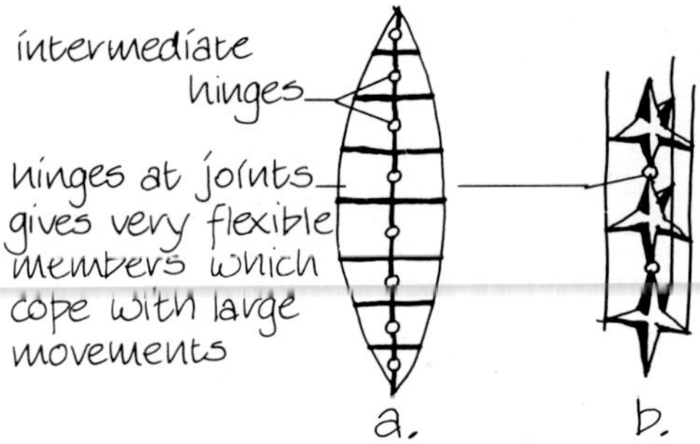

Fig. 155

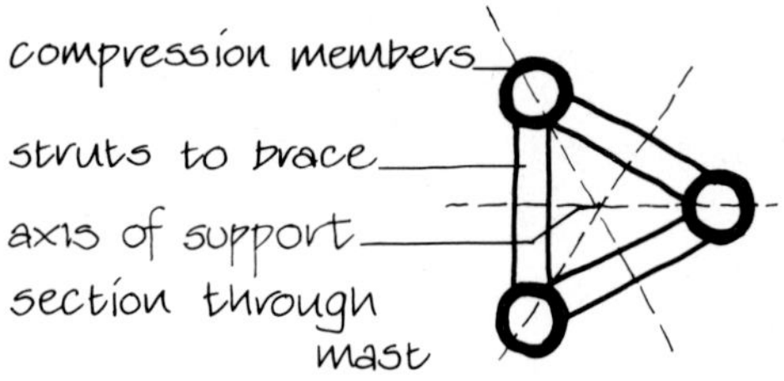

Fig. 156

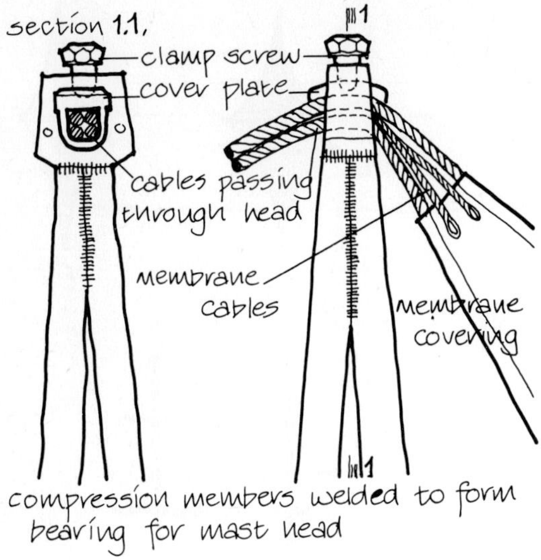


Fig. 157

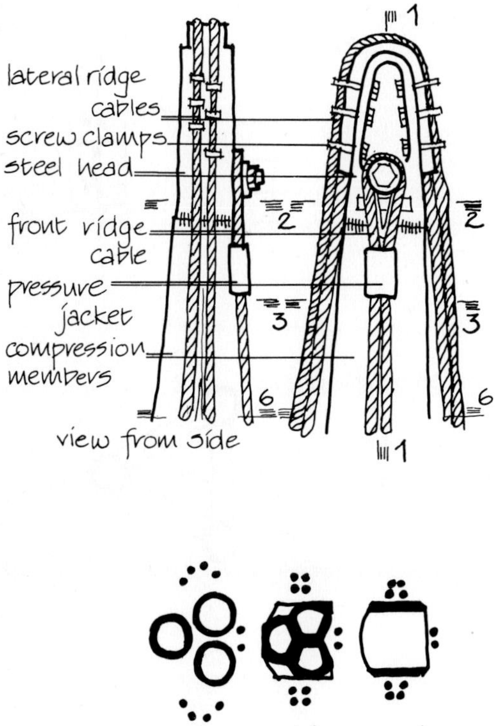


Fig. 158

The mast head shown in Fig. 157 has a bored-out steel unit with a clamp screw and cover plate for securing the cables and edge cables. The head is closed and this provides a bearing for the cables so that they can be deviated (see Appendix, Roof of dance floor, Cologne, p. 62).

The construction of a three-member support for a large membrane structure is shown in Fig. 158. Five pairs of cables are secured at the top of the pole; the lateral ridge cables are taken on a steel bearing unit and clamped on each side. The front ridge cables are separately bolted to the front

of the steel unit (Fig. 159) (see Appendix, Exhibition tents, Lausanne, p. 64).

Where the loads are not great very slim supports can be used, but all eccentric loads on the support must be avoided. If the lines of the carrying cables meet at differing angles they must intersect over the central axis of (Figs. 160, 161, 162) the support (see Appendix, Pavilions at Hamburg, p. 63).

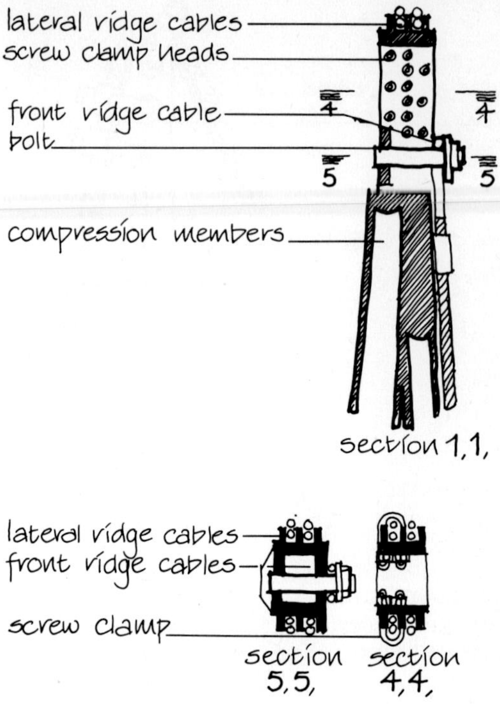


Fig. 159

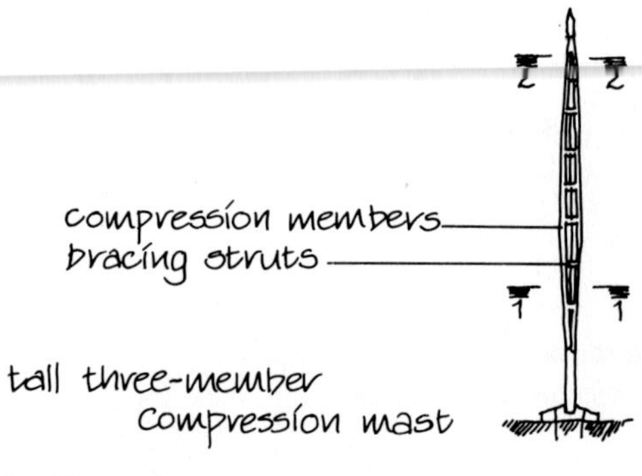


Fig. 160

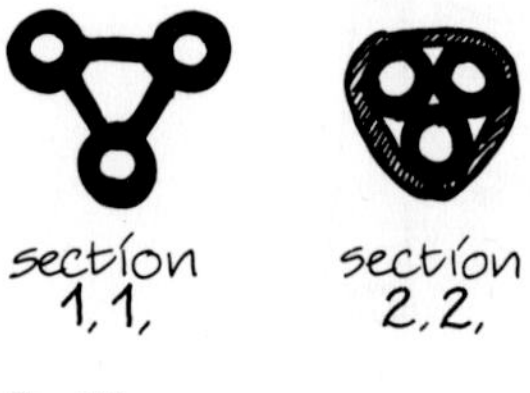


Fig. 161

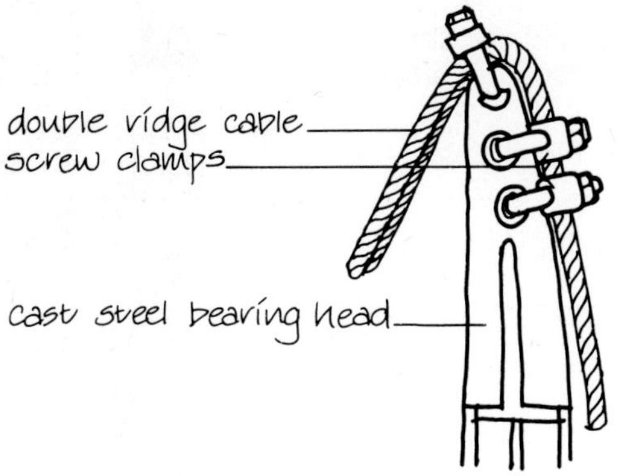


Fig. 162

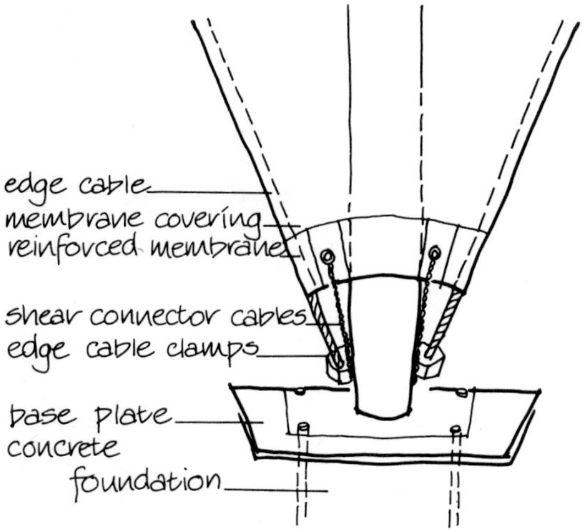


Fig. 163

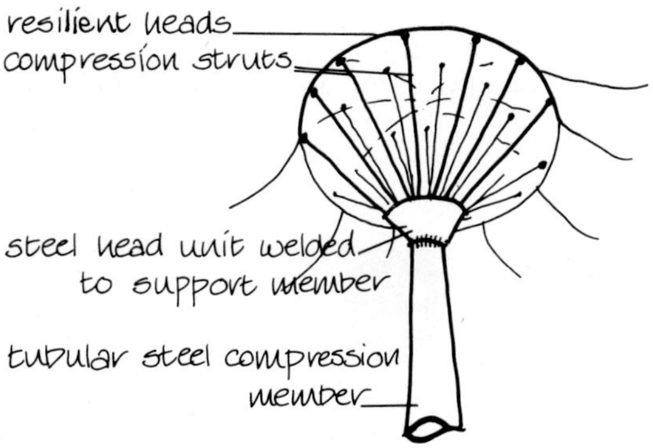


Fig. 164

The base of a support can include the attachment of edge cables; the cladding in this case is a membrane which is reinforced with several layers of material and secured using short shear resisting cables (Fig. 163).

The support can form either a single point or multiple points at the head of the mast. Using multiple points gives a humped peak where the forces are exerted on the net at predetermined points using struts with resilient heads. This method of construction of high points can be one of the cheapest and simplest, as it avoids complex cutting and attachments around the peak of the structure (Fig. 164).

## Non-structural considerations

### *Protection*

The positioning of the support, whether it is internal and therefore protected from the climatic conditions, or external and exposed to them, affects the amount of protection needed. Steelwork used externally obviously needs a stronger protective coating or more regular maintenance than any steelwork used internally.

Other materials which can be used to form supports need less or no maintenance. Laminated timber may need very little maintenance; concrete needs no maintenance.

### *Cost implications*

The cost of the masts will tend to constitute 10 to 15 per cent of the overall cost of the structure. Mast costs increase with increasing height and span of the structure, but so do other structural element costs, and the overall ratio of costs is maintained. When deciding on the type of support to be used, the height and loads are the critical factors. The most suitable material must then be chosen to suit the type of support required. For a steel mast the use of a compression strutted member with its material saving must be compared to the use of a hollow steel tapered tubular mast which is much simpler to manufacture for smaller loading conditions.

*Chapter 6*

# Claddings

Claddings cover the cable net without contributing to the structural performance. There are two main categories: continuous membranes including textile fabrics, metal and plastic sheets; and unit coverings, e.g. shingles, tiles, panel systems.

## Membranes

These are stretched skins achieved by cutting the material used to fit tightly over the cable network. They fall into two classes, the orthogonal anisotropic membranes, where the membrane acts in orthogonally (mutually perpendicular) preferential directions, that is, along the warp and weft. Sheet membranes are the second type. These are isotropic, that is, they have almost equal properties in all directions and exhibit the same behaviour in all directions under loading.

### *Orthogonal anisotropic membranes*

Organic textile fabrics such as cotton and linen canvas of strengths of 50–500 kg/m can be used for membrane coverings. These fabrics have a limited life of one to five years due to their natural deterioration. The length of life depends on the weight of the fabric used and the climatic conditions to which the membrane is subjected. These membranes are suitable only for short-term, non-protective structures because they cannot provide any control over the internal environment beyond protection from rain and wind.

Fabrics made from synthetic fibres of polyvinyl, polyacrylic and polyester have similar

weights and strengths to cotton fabrics but are cheaper to produce. However, they are very susceptible to deterioration under ultra-violet light and so have to be protected. This can be achieved by coating the fabric with pigmented plastic or metal foil, which considerably improves the life of the membrane but cuts down on the light transmission of the membrane. Like organic fibre fabrics, synthetic fibre fabrics are most suitable for small-span, short-term coverings.

Mineral fibres produce a fabric which does not elongate under stress, therefore it is essential that the pattern for the membrane is cut very accurately. This is a longer lasting fabric as it is not affected by ultra-violet light, but it is adversely affected by moisture; again, this membrane is most suitable for short-term claddings.

### *Isotropic membranes*

Plastic sheets of polyester, polyethylene or polyvinyl chloride have tensile strengths of 3–20 $kg/mm^2$ and can be coloured or translucent. Their strength under permanent stress is low so their life is limited to five years or so. Their environmental performance can be improved by using insulating layers of foamed plastic glued to the external sheet or enclosed by a lower sheet, or by introducing sealed air pockets over the surface of the membranes. Due to the lack of strength these membranes are suitable for small to medium span structures, such as swimming pool roofs, greenhouse coverings, etc.

### *Metal membranes*

Steel or aluminium are the strongest metals for use in sheets; they have strengths up to 90 $kg/mm^2$. The membrane is composed of small sections accurately cut and joined by welding, glueing or bolting and they are suitable for long-span, long-life structures with wide spaced cable networks.

When a structure is for long-term use it must be possible to control the internal environment; this can be achieved with the use of metal membranes. The factors to be considered are the lighting, heating and ventilation of the enclosed space. Natural lighting can be provided and

controlled by designing glazed openings in the membrane. To prevent heat loss, insulation can be applied by spraying or otherwise attaching a low-density material internally. Ventilation of the space may be natural, using the stack effect through a high point, or by mechanical means, depending on the occupancy and level of ventilation needed. However, it must be remembered that any metal covering needs to incorporate expansion joints to cope with the thermal movement, and venting positions to allow water vapour to escape.

### *Lattice sheets*

These are a combination of a sheet material with an orthogonally anisotropic fabric. The fabric is embedded in a sheet and acts as a reinforcement to give greater tensile strength, tearing strength and resistance to shear. Lattice sheets can be used in a similar manner to plastic sheets. However, they have a much longer life and so can be used for permanent structures. Their qualities of environmental control can be improved in the same manner as for plastic sheets.

The membranes must be connected firmly to the supporting cable network. Where especially high forces are acting, for instance over major cables and at peaks and anchorages where the membrane terminates, it is necessary to glue or weld-in reinforcing pieces.

The following give examples of membranes in use and the manner in which they can be attached to the cable net structure. Figure 165 shows a knot connection held within stitched or glued sleeves, which is then stitched or glued to the membrane to hold it in position. Due to the high cost of stitching, glueing is now the most commonly used method (see Appendix, Bandstand at Kassel, p. 61).

Where the membrane terminates at the mast head (Fig. 166), it is most highly stressed and susceptible to shear failure parallel to the edge. These forces are resisted by a short cable attached to a reinforcing disc attached to the strengthened edge of the membrane (see Appendix, Roof of dance floor, Cologne, p. 62).

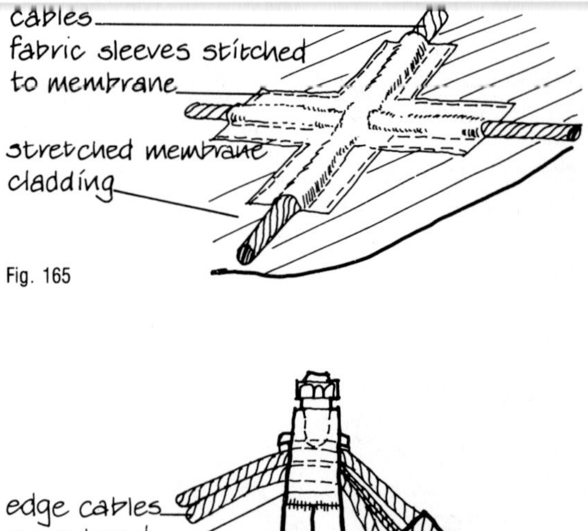


Fig. 165

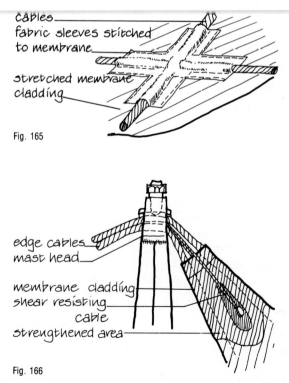

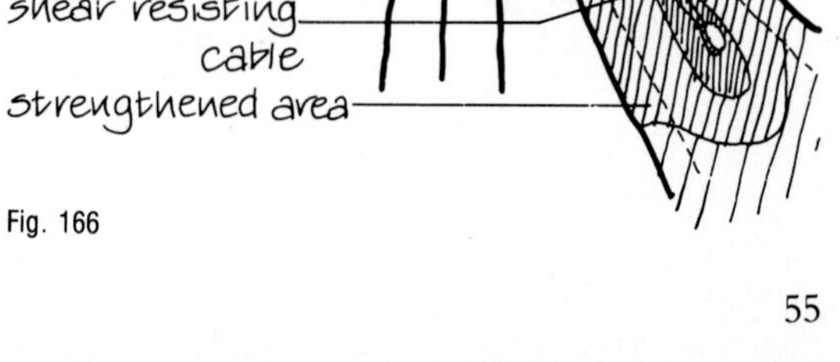


Fig. 166

THE LIBRARY
GUILDFORD COLLEGE
of Further and Higher Educatio

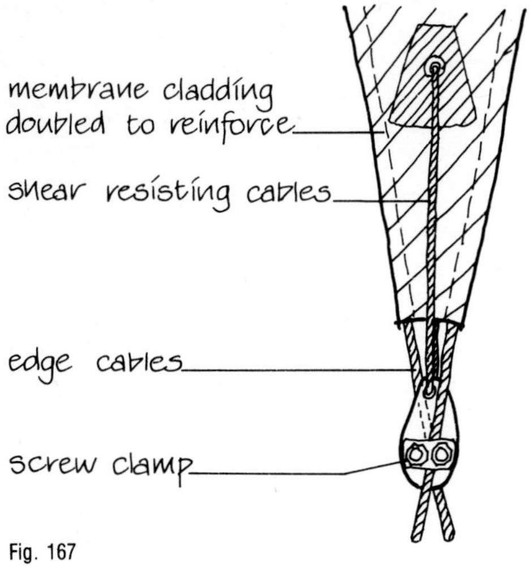


Fig. 167

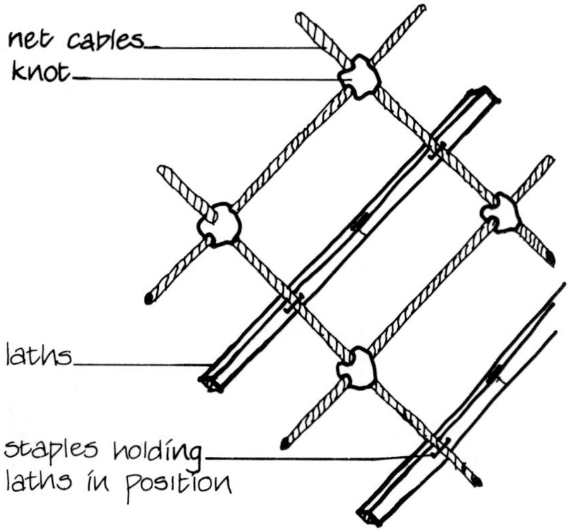


Fig. 168

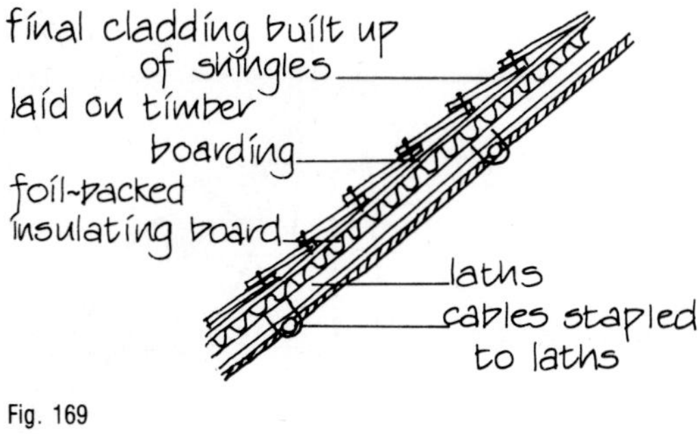


Fig. 169

In an anchorage where two cables converge at an acute angle (Fig. 167) the membrane is very highly stressed at the base. Therefore, it is attached to the anchorage with separate cables and reinforced over the bottom portion.

## Unit coverings

### *Small overlapping elements*

A cable net, once erected and tensioned, forms an ideal surface for using a cladding of small elements. This is because tiles or shingles can easily accommodate the curves of the surface without any special cutting. If the net cable spacing is between 400–600 mm it provides a structure which is easy for a man to walk about on, thus the laying of laths and the laying of any insulating boards and shingles or tiles is facilitated.

Figure 168 shows a method of laying laths between the cable meshes, which are then stapled in position. Short insulating boarding is applied in two directions with a metal foil on top: the shingles are then finally secured (Fig. 169) (see Appendix, I.L. Studio, Stuttgart, p. 66).

### *Panel systems*

As the cable network, due to its prestressed condition, is resisting all wind loads and needs no further bracing, the use of heavyweight cladding such as a conventional build up of boarding and shingle type waterproofing finish adds to the deadweight on the cables. This makes it necessary to increase the size of the cables to cope with the larger forces, which increases the cost of the cable network. The use of a lightweight covering can be more economical as it does not impose extra weight on the cable structure.

A light panelling system which is made up of rigid individual elements has to cope with the curvature of the structure. The panels must either be cut to the form of the structure or jointed in a flexible manner so that they will conform to the shape of the structure once laid over the surface. Panels of plastics, which can be transparent, translucent or opaque, or have insulation sandwiched between two layers, are suitable for use. Timber, fibreboard and lightweight concrete are also appropriate as they are also relatively light in weight.

Panel systems give the structure a durable finish and are therefore applicable to permanent structures. In an enclosed long-life building the internal environment must be regulated. Plastic panels such as 'Plexiglas' give great flexibility of daylighting. The whole building surface can be transparent or translucent or portions can be designed to be transparent within an overall opaque surface.

An inherent problem of lightweight construction when the enclosed volume is being heated is that of heat loss. To prevent the thermal transmission of heat it is necessary to heavily insulate the cladding.

The following show two methods of applying cladding panels and how the panels may be connected to each other. One method is to keep the insulation separate from the panel by suspending the insulation as an independent layer below the cladding (Fig. 170). The insulation can be light boarding 100–150 mm thick sandwiched between two lattice sheet membranes. The panels are laced together before hoisting them into position. They may then be secured using trefoils suspended from the main cable network. Any light insulation boarding or extruded insulation can be suspended to form the insulating ceiling (Fig. 171).

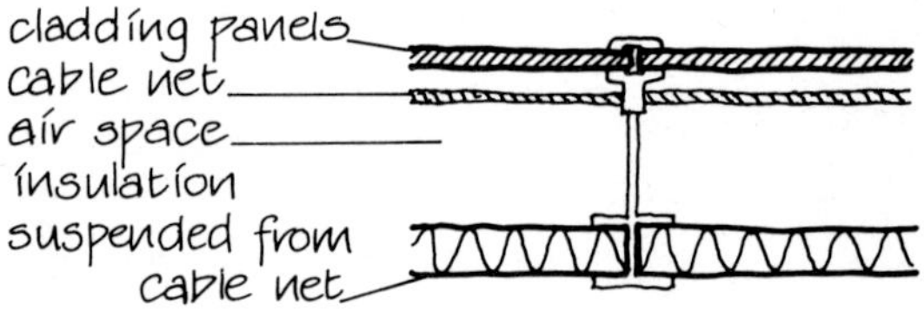


Fig. 170

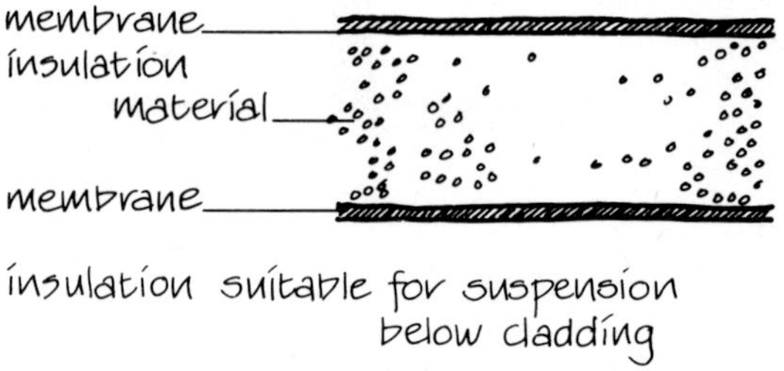


Fig. 171

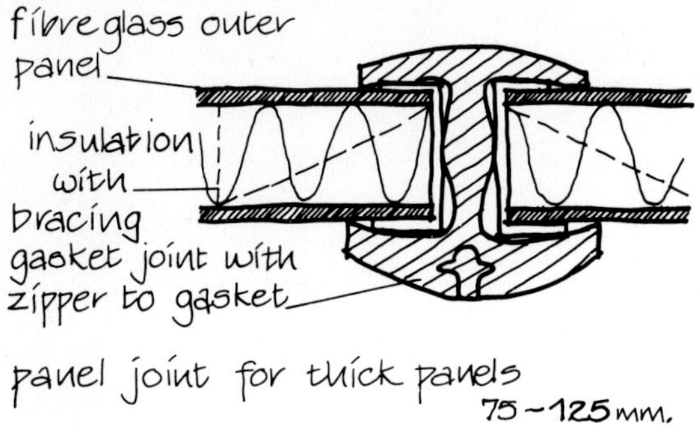


Fig. 172

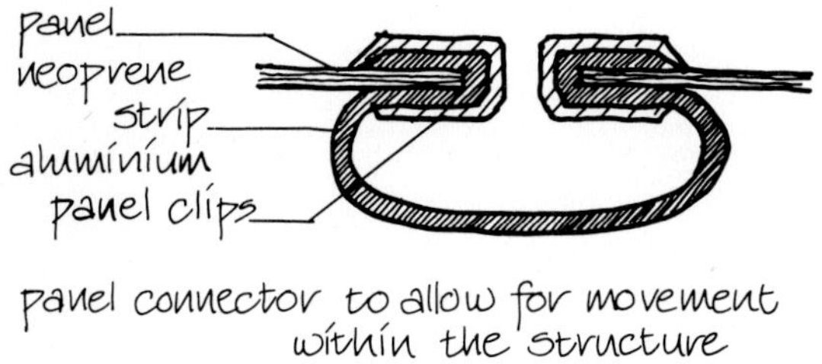


Fig. 173

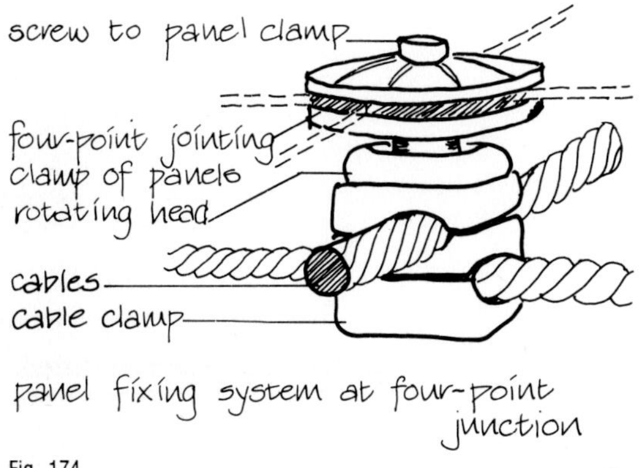


Fig. 174

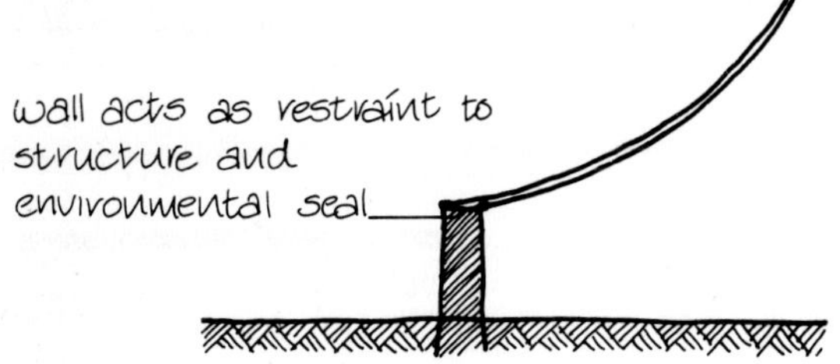


Fig. 175

Alternatively, the cladding panel can form a sandwich, with the insulation within the panel itself (Fig. 172). Such panels are commercially available with 'U' values of 0.5 watts/$m^2$ per deg. C or less depending on the thickness of the panel.

Movement between panels can be catered for by the separation joint shown in Fig. 173.

For panels which are fixed above a cable net, the four-point jointing clamp can be used (Fig. 174).

## Junctions

When the structure is permanent and it is necessary to have a weatherproof boundary which forms an environmental seal between the net edge and the ground then a wall can be used which can act as the restraint and anchorage for the structure (Fig. 175). The junction between the net and

the wall can be rigid (Fig. 176) or the wall can be a non-loadbearing infill. In this case the junction must be capable of accepting thermal movement in the roof structure (Fig. 177).

The other factor which affects the junction is the radius of curvature of the net boundary. The junction must accommodate the curve between the net and the wall (Figs. 178 and 179).

The recent development of resilient neoprene gaskets and thiokol-based mastic compounds which do not deteriorate with age has made the production of flexible junctions possible.

Figure 180 demonstrates a junction of a panel type infill between uprights and a cable network. The mullion is hollow and attached to the edge cable with a thin cable; this gives the upright flexibility. The panel system is connected to the edge cable with a pliable gasket.

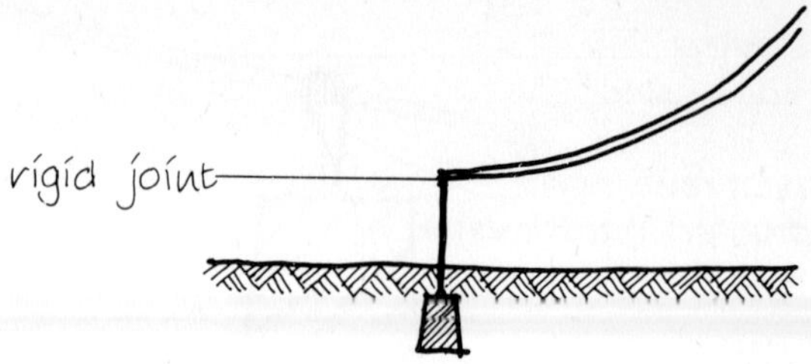


Fig. 176

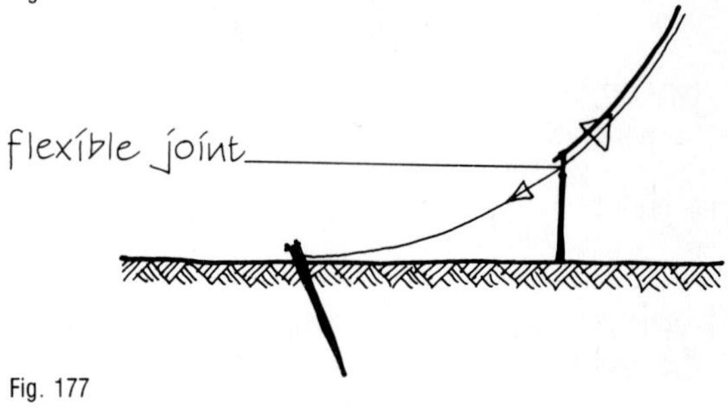


Fig. 177

Fig. 178

Fig. 179

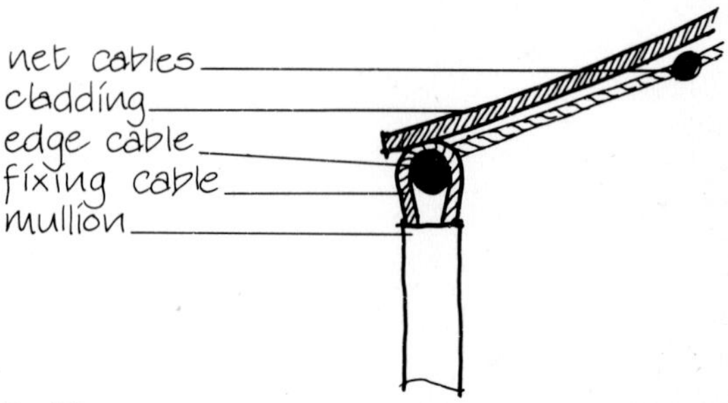


Fig. 180

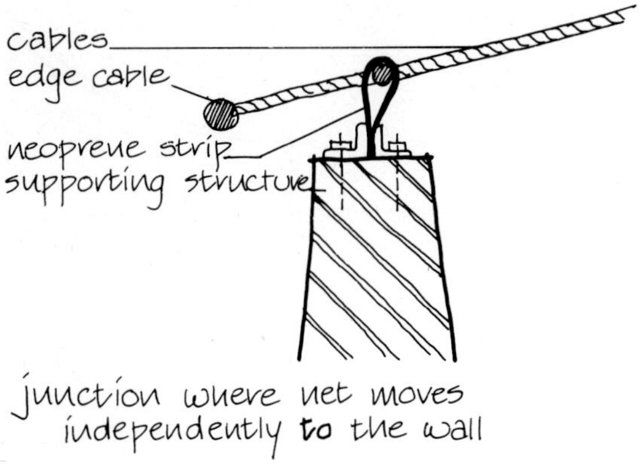


Fig. 181

typical value of linear thermal expansion of mild steel

$= 12\ \text{mm/mm}\ 10^{-6}$ per °C

temperature change

$= 15\ °C$

length of cable

$= 25 \times 10^{3}\ \text{mm}$

elongation of cable

$$'e' = 15 \times 25 \times 10 \times 12\ \text{mm}$$

$$'e' = 4.5\ \text{mm}$$

Fig. 182

A yielding junction between a solid infill and a net cable must allow the structure to move up and down. The detail in Fig. 181 shows a neoprene strip folded over a perimeter cable and held by a metal plate.

When the infill becomes loadbearing and acts as a support or an anchorage the connection would be designed as a knot element as discussed previously.

## Non-structural considerations

### *Thermal movement*

To enable the designer to calculate the elongation of a cable under a temperature change manufacturers provide specifications which include the linear thermal expansion in mm/mm$^{10-6}$ per deg. C. For instance, under a temperature variation from 5 °C to 20 °C, a possible change in temperature during summer, the elongation over a cable length of 25 m will be 4.5 mm (Fig. 182).

The cladding will tend to be thermally dissimilar to the cable thus there will be differential movement between the cable net and the cladding system. To accommodate this movement the cladding must be capable of sliding at any fixed points with the cable net and at any junctions between rigid uprights and the cable net. This can be achieved by using either sliding clamps at fixed points, where the cladding is allowed to move independently, or by incorporating expansion joints within the cladding which will allow the cladding to expand.

### *Cost implications*

The cost of the cladding is largely dependent upon the degree of environmental control which it is required to achieve. If the structure is lightweight and short term – for instance, a canopy structure which is not intended to enclose the space but to provide shelter – then a cheap cladding of a fabric, sheet or lattice sheet membrane can be utilised.

In larger permanent structures the percentage cost of the cladding will be higher because the materials being used are more expensive and the methods of connecting the cladding to the cable net are more complex.

# Appendix

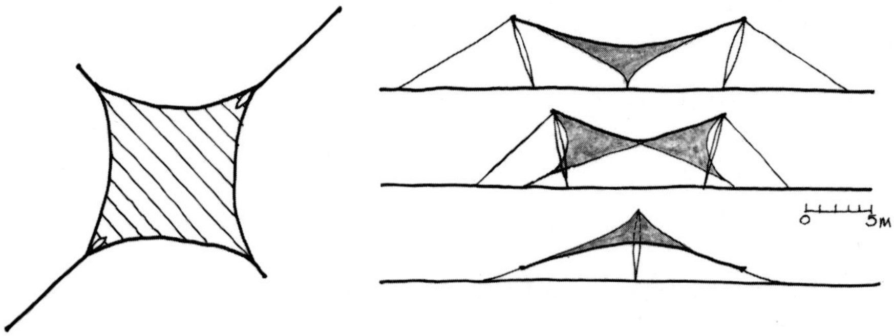

Fig. 183

***1955 Bandstand for the Federal Garden Exhibition, Kassel, West Germany***

*Construction* Two high points, two low points. Cotton membrane stressed over parallel cables. Cables sewn into pockets to strengthen membrane. Tapered steel masts 5 m. 16 mm diam. edge cables. Buried concrete gravity anchors to resist up to 22 tonnes tensile force.

*Architect/Designer: Frei Otto. (Overall direction: H. Mattern: collaborator, S. Lohs).*

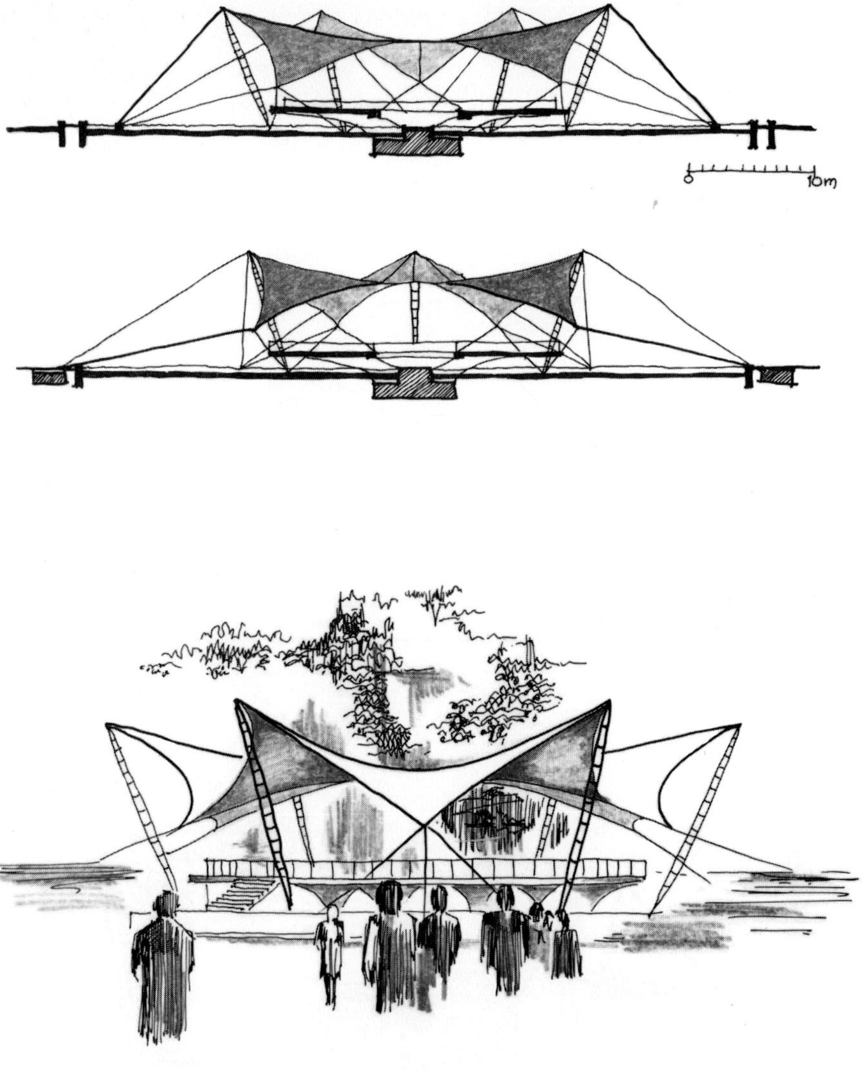


Fig. 184

---

***1957 Roof for dance floor, Federal Garden Exhibition, Cologne, Germany***

*Construction* Six high points, six low points. High points supported by compression-trussed masts. Low points secured by anchor cable to concrete gravity base. Edge cables 22 mm diam. with 120 kg/m prestress to membrane. Masts 10.4 m. Span 33 m. Stressed membrane, 1 mm thick, cotton, tearing strength (weft direction) 2,790 kg/m.

*Architect/Designer: Frei Otto. (Collaborators: E. Bubner, S. Lohs, D.R. Frank).*

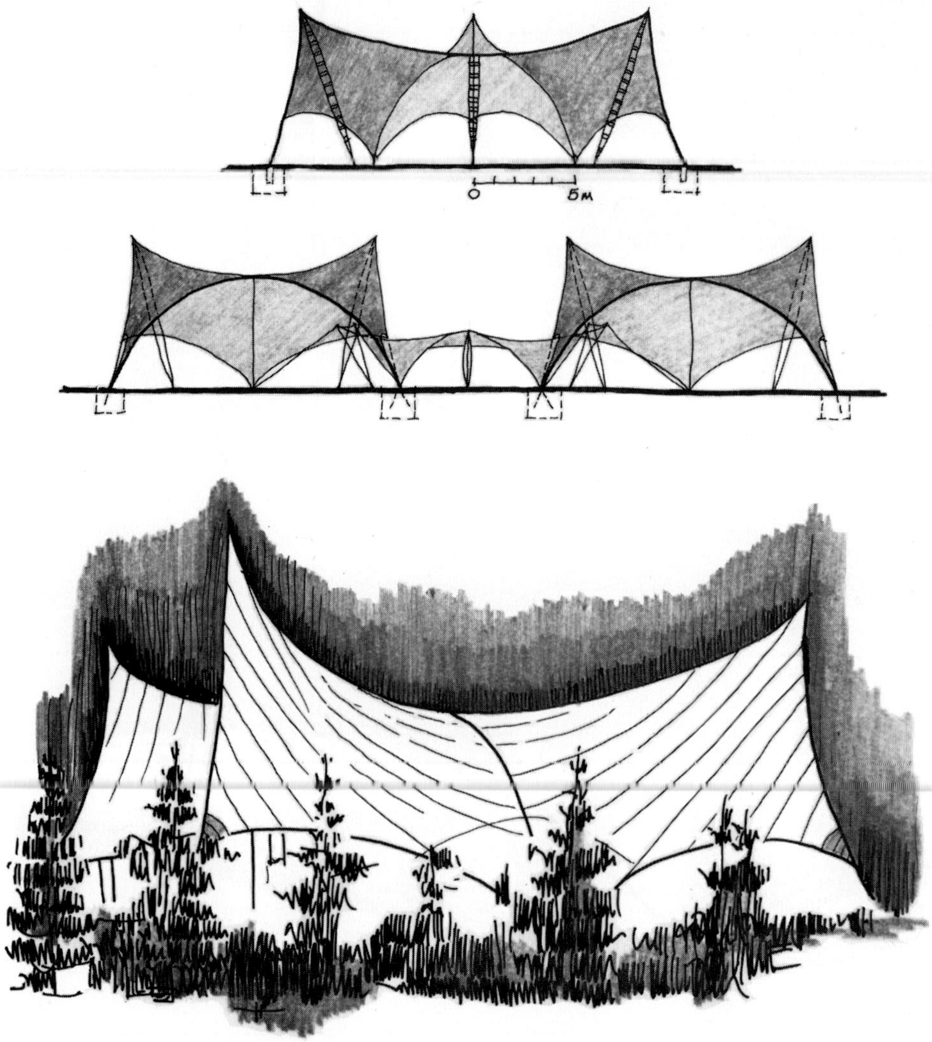


Fig. 185

***1963 Pavilions at the International Horticultural Exhibition, Hamburg***

*Construction* Star-shaped undulating system. Four high points, four low points. Membrane stressed over parallel cables. Compression trussed masts, 18 m high. Span diagonally between masts, 18 m. Membrane cotton fabric.

*Architect/Designer: Frei Otto, Hans Hinrich Haberman, Christian Hertling, John Koch.*

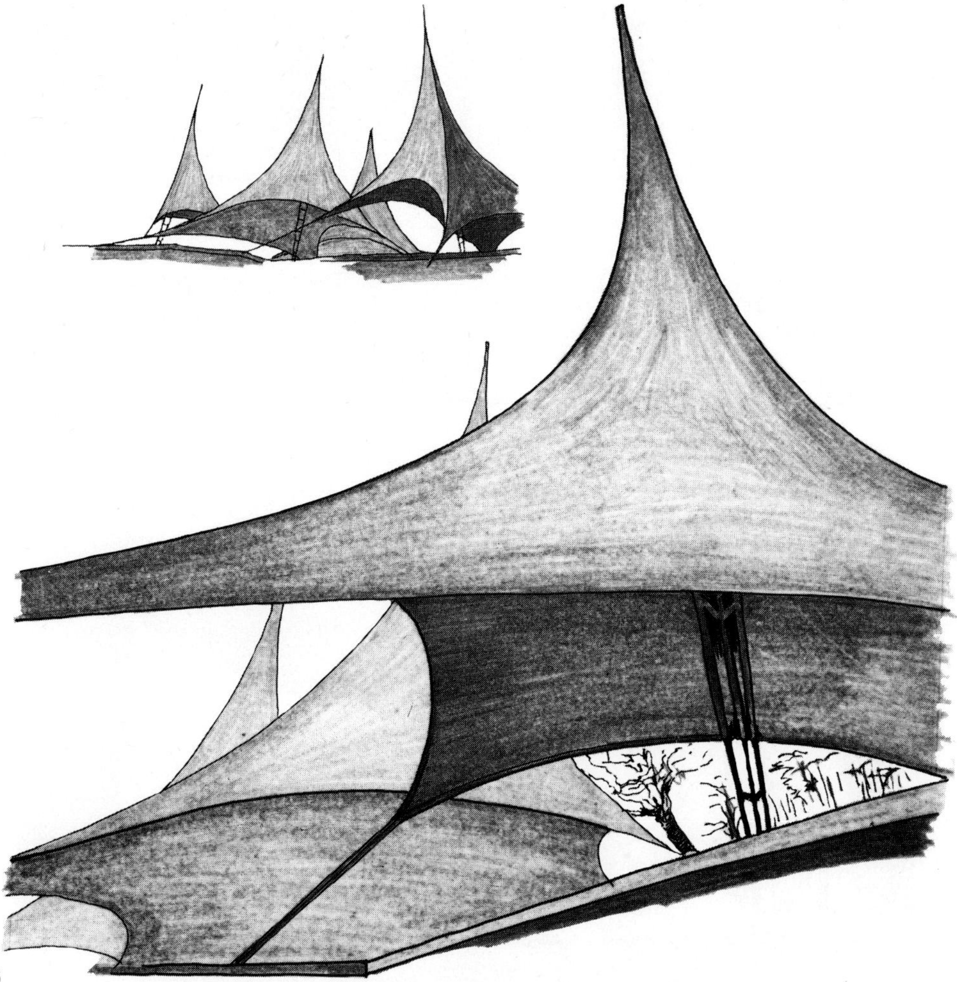

Fig. 186

***1964 Tents at Lausanne for the Swiss National Exhibition. Designed as a sculptural representation of the Swiss Alps***

*Construction* Edge cables, four-point surface – each individual tent supported by a compression trussed mast. Rectangular cable network with membrane cladding. Edge cables 22 mm diam. steel cable. Net cables 8 mm diam. steel cables. Mast 24 m. Span 36 m.

*Architect/Designer: Marc Saugey. Consultant – Frei Otto.*

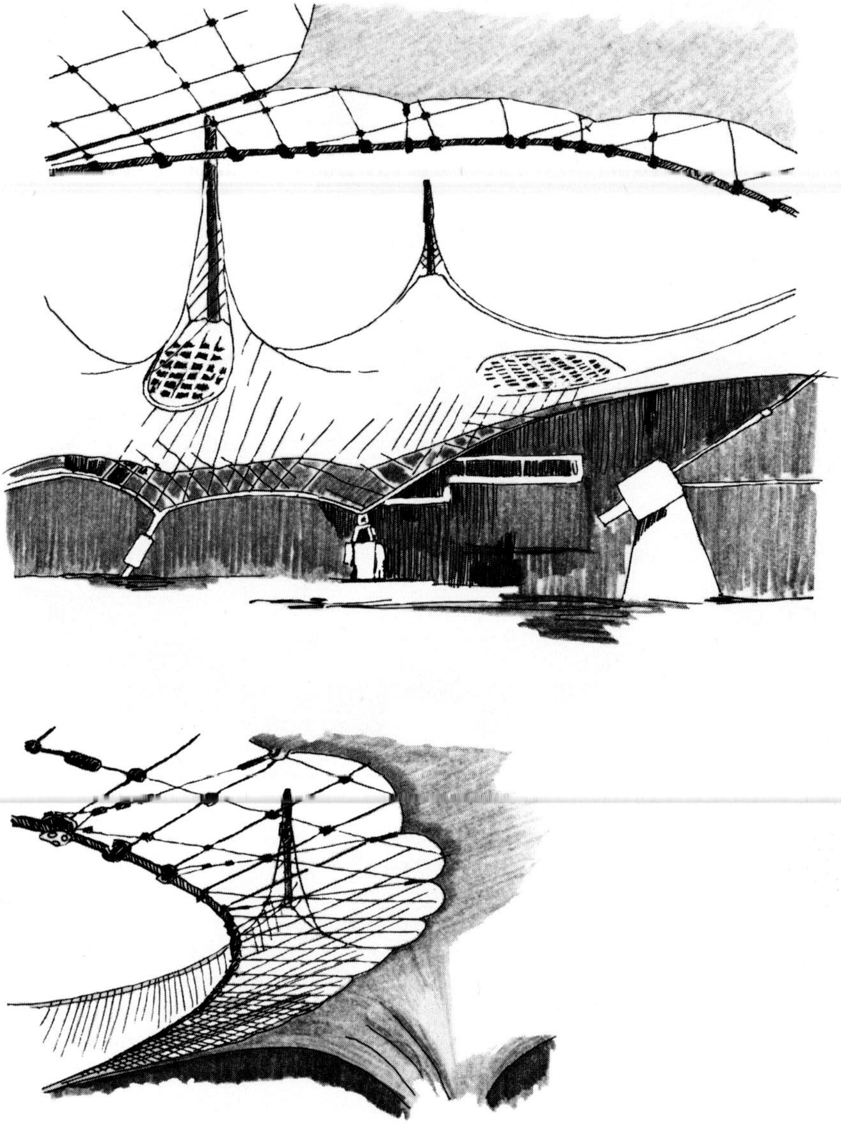

Fig. 187

***1967 German Pavilion for Expo '67 Montreal, Canada***

*Construction* Cable network, supported on tubular hollow steel mast. Anchors using anchor needles. Cladding, suspended below cable network of a pvc membrane. Maximum length of span 130 m. Maximum width of span 105 m. Masts vary from 14–38 m. Net cables 12 mm diam. steel. Edge cables 54 mm diam. steel. PVC membrane coated with ultraviolet absorber.

*Architect/Designer: Rolf Gutbrod, with Kiess, Kendel, Medlin and Frei Otto*

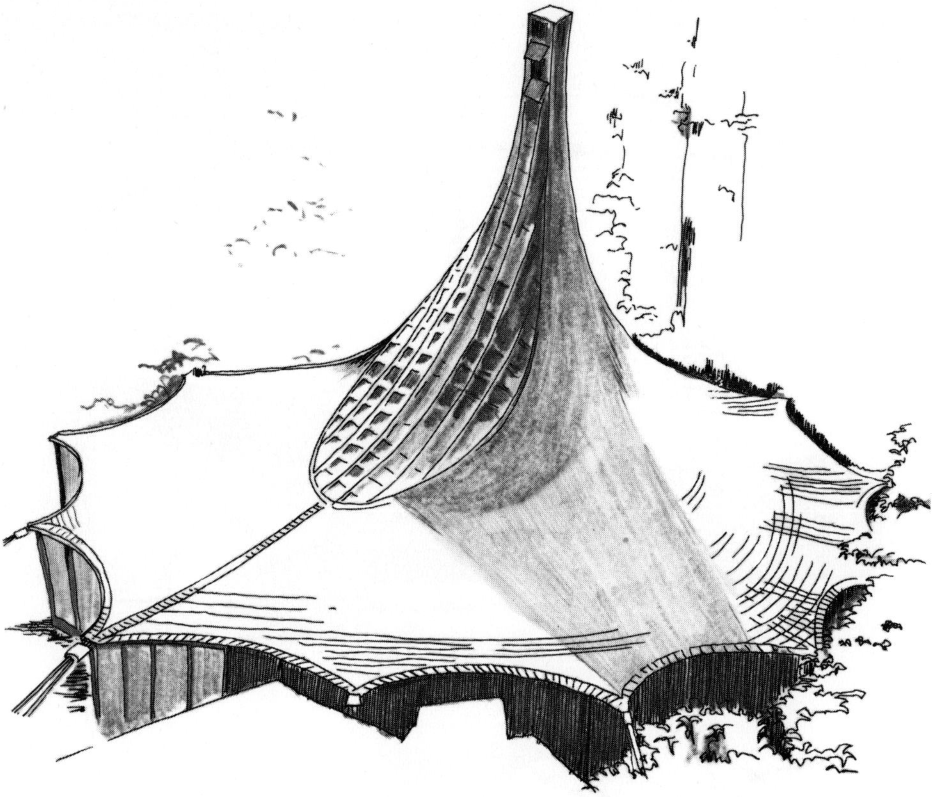

Fig. 188

***1968 Studio for the Institute of Lightweight Structures, Stuttgart***

*Construction* Cable network, one high point. Twelve anchorages. Glazed gathering (eye) loop. Cladding of Eternit asbestos shingles on soft wood laths on mineral wool insulation. Cable net 12 mm diam. at mesh width of 500 mm. Edge cable 54 mm diam. Mast tapered hollow tubular steel maximum diameter 500 mm. Height 17 m.

*Architect/Designer: B. Buckhardt, F. Otto with F. Kugel, G. Minke, B. Rasch.*

Fig. 189

***1972 Stadium roofs for main sports arena, athletics arena and swimming hall for 20th Olympic Games, Munich, West Germany (Fig. 189)***

*Construction* Cable net with Plexiglas panel cladding above cables. Roof area athletics stadium 34,500 $m^2$. Sports arena 21,750 $m^2$. Swimming hall 11,900 $m^2$. Masts 50–80 m hollow steel tubular masts. Cable nets 12 mm diam. steel. Edge cable 82 mm diam. steel. Width of mesh 600 mm. Plexiglas panels 1,800 × 1,200 mm.

*Architect/Designer: Behnish and Partners with Frei Otto and F. Bubner.*

# Further reading

For a general overview of structural form we would recommend:

Drew, Philip, *Frei Otto, Form and Structure*, Granada, 1976.
Drew, Philip, *Tensile Architecture*, Granada, 1979.
Roland, Conrad, Frei Otto, Structures, Longman, 1970.
Otto, Frei (Ed.), Tensile Structures, M.I.T. Press, 1969.

And for a straightforward treatment of structural analysis:

Krishna, Prem, Cable Suspended Roofs, McGraw-Hill, 1978.

The Institute of Lightweight Structures, Stuttgart has produced a number of books in this subject area, the most directly relevant being:

I.L.1. *The Experimental Determination of Minimal Nets.*
I.L.5. *Convertible Roofs.*
I.L.8. *Nets in Nature and Technics.*

# Definition of terms

The following terms are in general use when discussing cable net structures:

*Cable*
A flexible, tension-resistant element usually made from metallic fibres: steel, stainless steel, nickel, copper, molybdenum. They can also be constructed of man-made fibres such as nylon, carbon fibres, etc.

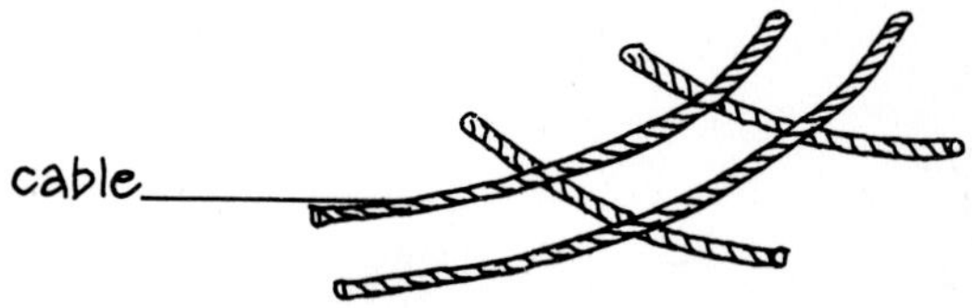


*Cable chord*
An imaginary straight line connecting the points of support of a hanging cable.

*Net*
A tensile-stressed mesh structure, consisting of cables and nodes which form the open flexible system or mesh.

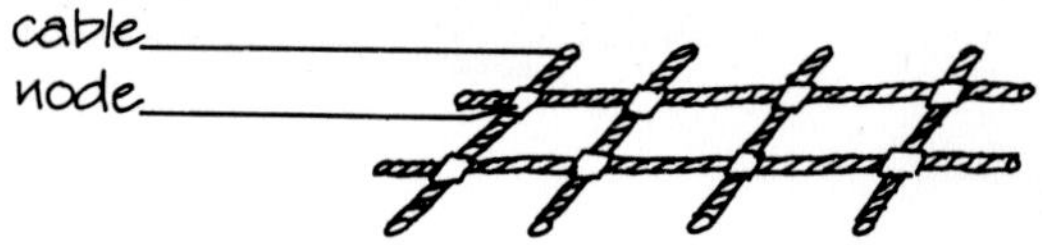


*Synclastic curvature*
When a structure is unilaterally curved in a dome shape it is said to be synclastically curved.

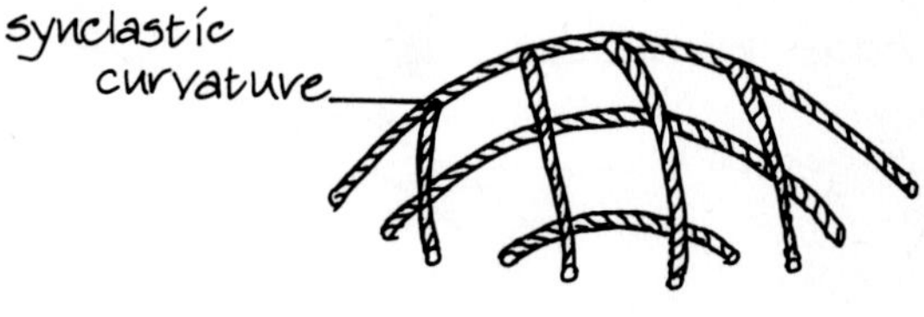


*Anticlastic curvature*
When a mesh is unilaterally curved in a saddle shape it said to be anticlastically curved.

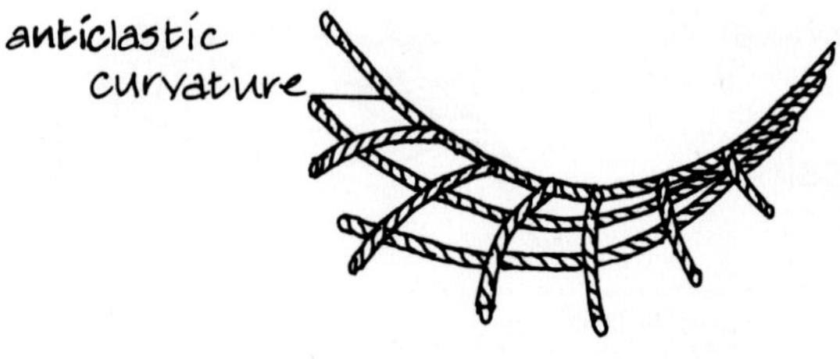

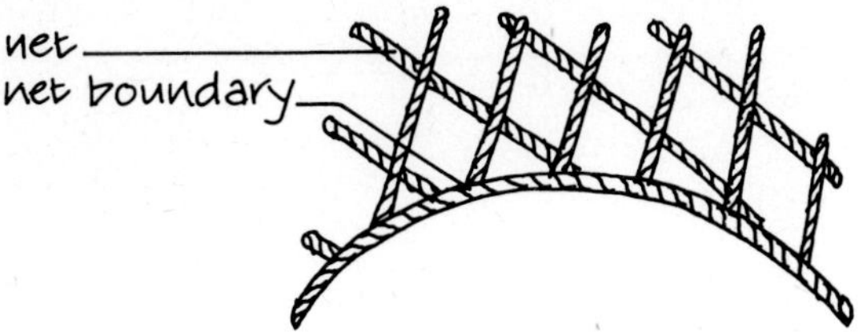


*Net boundary*
A closed, spatially curved limitation line which totally encloses the geometric form of the net.

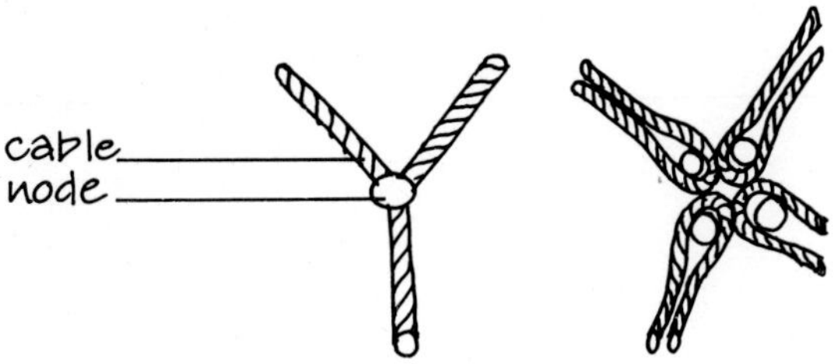


*Node*
A node is a point at which two or more cables meet and where the cable can end or continue through the node.

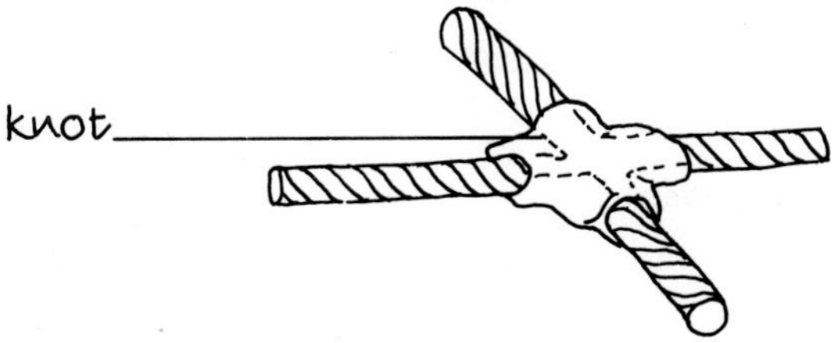


*Knot*
A knot is the building component which forms a node.

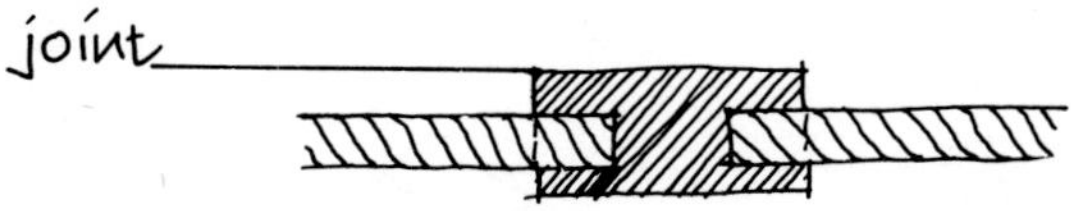


*Joint*
A joint is a building component by means of which two cables are joined longitudinally.

*Edge knot*
An edge knot occurs at the point where a cable forming part of the net is connected to an edge cable.

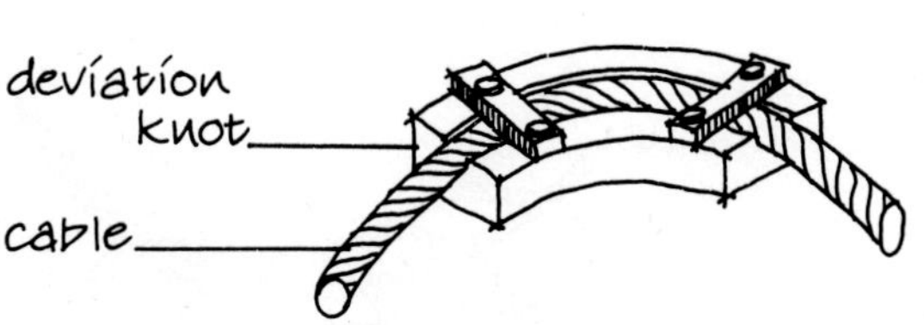


*Deviation knot*
A deviation knot is the building component by means of which the direction of a continuous cable is constantly changed along a predetermined flat or spatial curve.

*Branching knot*
A branching knot has the same function as a deviation knot but instead of the cable or cables being continuous through the knot they end within it.

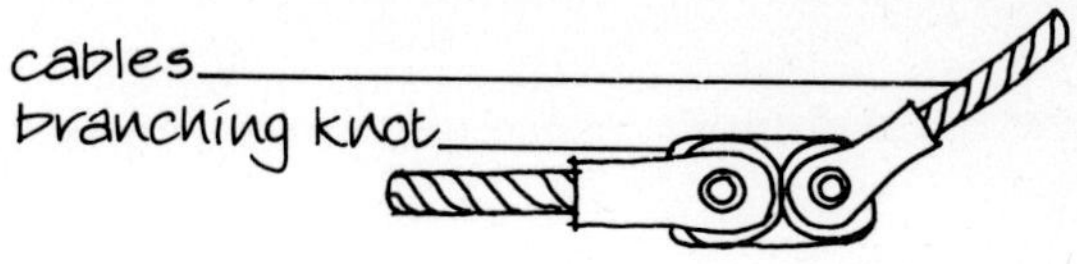


*Terminal*
A terminal is the building component which ends the cable and is used to attach the cable to other building elements.

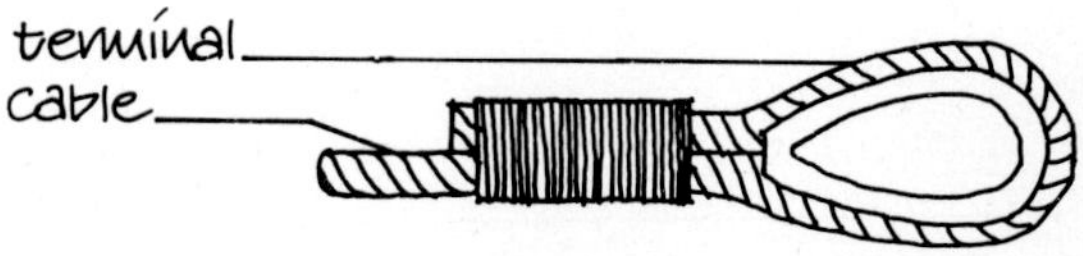


*Anchorage*
An anchorage is the means by which tensile forces are transferred from the net cables at one point or several points into the ground.

*Net support*
A net support is the structural element which supports the net and fixes its position in space in relation to the ground.

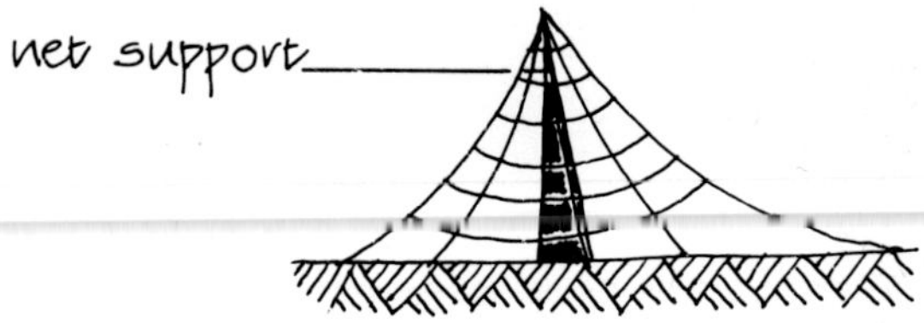


*Mesh width*
The mesh width is the distance between the node points of the cable network.

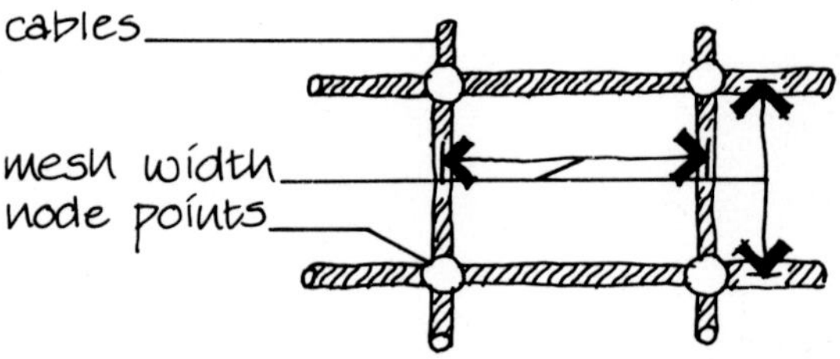

WITHDRAWN

Guildford College
Learning Resource Centre

Please return on or before the last date shown
This item may be renewed by telephone unless overdue

| | | |
|---|---|---|
| 07 JUN 2006<br>28 JUN 2006 | | |

Class: 721.042 CHA

Title: The Technology of Suspended Cable Net Structures

Author: CHAPLIN Frances

141303

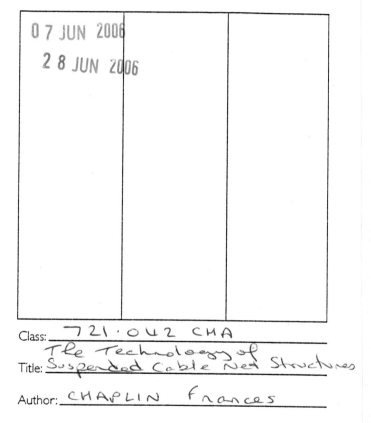